MAINTENANCE SCHEDULING
IN
RESTRUCTURED POWER SYSTEMS

THE KLUWER INTERNATIONAL SERIES IN ENGINEERING AND COMPUTER SCIENCE

Power Electronics and Power Systems

Series Editor

M. A. Pai

Other books in the series:

POWER SYSTEM OSCILLATIONS
Graham Rogers, ISBN: 0-7923-7712-5
STATE ESTIMATION IN ELECTRIC POWER SYSTEMS: *A Generalized Approach*
A. Monticelli, ISBN: 0-7923-8519-5
COMPUTATIONAL AUCTION MECHANISMS FOR RESTRUCTURED POWER INDUSTRY OPERATIONS
Gerald B. Sheblé, ISBN: 0-7923-8475-X
ANALYSIS OF SUBSYNCHRONOUS RESONANCE IN POWER SYSTEMS
K.R. Padiyar, ISBN: 0-7923-8319-2
POWER SYSTEMS RESTRUCTURING: *Engineering and Economics*
Marija Ilic, Francisco Galiana, and Lester Fink, ISBN: 0-7923-8163-7
CRYOGENIC OPERATION OF SILICON POWER DEVICES
Ranbir Singh and B. Jayant Baliga, ISBN: 0-7923-8157-2
VOLTAGE STABILITY OF ELECTRIC POWER SYSTEMS, Thierry Van Cutsem and Costas Vournas, ISBN: 0-7923-8139-4
AUTOMATIC LEARNING TECHNIQUES IN POWER SYSTEMS, Louis A. Wehenkel, ISBN: 0-7923-8068-1
ENERGY FUNCTION ANALYSIS FOR POWER SYSTEM STABILITY, M. A. Pai, ISBN: 0-7923-9035-0
ELECTROMAGNETIC MODELLING OF POWER ELECTRONIC CONVERTERS, J. A. Ferreira, ISBN: 0-7923-9034-2
MODERN POWER SYSTEMS CONTROL AND OPERATION, A. S. Debs, ISBN: 0-89838-265-3
RELIABILITY ASSESSMENT OF LARGE ELECTRIC POWER SYSTEMS, R. Billington, R. N. Allan, ISBN: 0-89838-266-1
SPOT PRICING OF ELECTRICITY, F. C. Schweppe, M. C. Caramanis, R. D. Tabors, R. E. Bohn, ISBN: 0-89838-260-2
INDUSTRIAL ENERGY MANAGEMENT: *Principles and Applications*, Giovanni Petrecca, ISBN: 0-7923-9305-8
THE FIELD ORIENTATION PRINCIPLE IN CONTROL OF INDUCTION MOTORS, Andrzej M. Trzynadlowski, ISBN: 0-7923-9420-8
FINITE ELEMENT ANALYSIS OF ELECTRICAL MACHINES, S. J. Salon, ISBN: 0-7923-9594-8

MAINTENANCE SCHEDULING IN RESTRUCTURED POWER SYSTEMS

M. SHAHIDEHPOUR, PhD
ILLINOIS INSTITUTE OF TECHNOLOGY
CHICAGO, ILLINOIS

M. MARWALI, PhD
ABB ENERGY INFORMATION SYSTEMS
SANTA CLARA, CALIFORNIA

KLUWER ACADEMIC PUBLISHERS
NORWELL, MASSACHUSETTS

Distributors for North, Central and South America:
Kluwer Academic Publishers
101 Philip Drive
Assinippi Park
Norwell, Massachusetts 02061 USA
Telephone (781) 871-6600
Fax (781) 871-6528
E-Mail <kluwer@wkap.com>

Distributors for all other countries:
Kluwer Academic Publishers Group
Distribution Centre
Post Office Box 322
3300 AH Dordrecht, THE NETHERLANDS
Telephone 31 78 6392 392
Fax 31 78 6546 474
E-Mail services@wkap.nl>

Electronic Services <http://www.wkap.nl>

Library of Congress Cataloging-in-Publication

Shahidehpour, M., 1955-
Maintenance scheduling in restructured power systems / M. Shahidehpour, M. Marwali.
p. cm. -- (The Kluwer international series in engineering and computer science ; SECS 562. Power electronics and power systems)
Includes bibliographical references and index.
ISBN 0-7923-7872-5 (alk. paper)
1. Electric power systems--Maintenance and repair. 2. Production scheduling. I. Marwali, M. II. Title. III. Kluwer international series in engineering and computer science ; SECS 562. IV. Kluwer international series in engineering and computer science. Power electronics & power systems.

TK1005 .S445 2000
621.31'2--dc21

00-031338

Printed on acid-free paper.

Printed in the United States of America

To the memory of my father who taught me the value of education.

M. Shahidehpour

TABLE OF CONTENTS

LIST OF FIGURES

Figure

LIST OF TABLES

Table

Table

Table

LIST OF SYMBOLS

Symbol	Definition
A	Fit parameter for PV (between 1 and 5)
A_{km}^{i}	Sensitivity coefficient of unit i to flow of line k
$\boldsymbol{B}$	Susceptance matrix
c_{it}	Lost of revenue per-MW at bus i in week t due to real power interruption; in hour is $c_{i\tau}$
$c'_{i\tau}$	Lost of revenue per-MVAR at bus i in hour τ due to reactive power interruption or purchasing reactive power
c_{mt}	Fuel cost per MBtu of fuel m in week t
$C(\tau)$	State of charge of battery at hour τ
C_o	Coefficient of saturation current density $[V^{-1}]$
C_{01}	Coefficient $[Am^{-2}K^{-3}]$
C_1	Coefficient $[Am^{-2}K^{-5/2}]$
C_f	State of charge of battery at the end of study horizon
C_{it}	Generation maintenance cost for unit i in week t
C'_{kt}	Transmission maintenance cost per-line in right-away k in week t; in hour is $C'_{k\tau}$
C_n	Maximum state of charge of battery
Cp_i	Capacity of thermal unit i
C_s	Battery state of charge at beginning of study horizon
$\underline{C}$	Minimum state of charge of battery
d_i	Duration of maintenance for generator unit i
$d_{i\tau}$	Load demand at bus i at hour τ
d'_k	Duration of maintenance for line k
$\boldsymbol{d}$	A vector of the weekly peak load at every bus in week t
DR_i	Ramp down rate of unit i (MW/hr)
e_i	Earliest period to begin maintenance of generator unit i
e'_k	Earliest period to begin maintenance of line k

E	Expected value
$\boldsymbol{E}^{(i)}(J)$	Energy in segment J after unit i committed
$E^{i}_{km}(j)$	Sensitivity coefficient matrix of unit i to flow of line k due to outage of line j
E_{Di}	Unserved energy after the first i units committed
E_{Gi}	Expected generation for unit i (i=1 is PV)
EMA	Emission cap for area emission in the study period
EMA_t	Emission cap for area emission in week t
EMS	Emission cap for system emission in the study period
EMS_t	Emission cap for system emission in week t
$\boldsymbol{f}$	Power flow at peak load in vector form
$\bar{\boldsymbol{f}}$	Upper limit of line flow in vector form
$\underline{\boldsymbol{f}}$	Lower limit of line flow in vector form
$f_{k\tau}$	Flow of line k at hour τ
$\overline{f}_k$	Upper limit flow of line k
$\underline{f}_k$	Lower limit flow of line k
f_{ci}	Fuel cost of unit i
$\boldsymbol{f}_{pv}$	Power density function for the PV
$\boldsymbol{f}^{\boldsymbol{c}}(j)$	Flow limit vector for line outage j
$\boldsymbol{f}^{\boldsymbol{s}}$	Steady state flow limit in vector form
$F_i(g_{i\tau})$	Fuel cost of unit i when generating power is $g_{i\tau}$
$\boldsymbol{F}^{\boldsymbol{c}}(j)$	Penalty vector for contingency flow constraints in case of line outage j
$\boldsymbol{F}^{\boldsymbol{s}}$	Penalty vector for steady state flow constraints
FL	Lower limit of fuel constraint for a group of units
FU	Upper limit of fuel constraint for a group
$\boldsymbol{g}$	Vector of power generation by each unit at peak load in week t
$\overline{\boldsymbol{g}}$	Maximum generation capacity in vector form

$\overline{g}_i$	Upper real power limit of unit i; in BUC: maximum limit for power to supply load and bilateral contracts
$\underline{g}_i$	Lower real power limit of unit i; in PBUC: minimum limit for power to supply load and bilateral contracts
$g_{i\tau}$	Real power output of unit i at hour τ, PBUC: power to supply bilateral contracts and to be sold to supply load from unit i at hour τ
$g_{i\tau}^{o}$	Bilateral contracts for energy to supply load from generator i at hour τ
$\boldsymbol{g}^{c}(j)$	Reactive limit vector for line outage j
$\boldsymbol{g}^{s}$	Steady state reactive limit vector
G_h	Global radiation on a horizontal plane (W/m^2)
G_{oh}	Global radiation outside the atmosphere (W/m^2)
G_t	Instantaneous radiation on a tilted surface (W/m^2)
$\boldsymbol{G}^{c}(j)$	Penalty vector for voltage constraints in line outage j
$\boldsymbol{G}^{s}$	Penalty vector for steady state voltage constraints
h_{im}	Average heat rate (MBtu/MWh) of unit i burning fuel m
h_k	Duration hour of maintenance for lines in right-away k
H_i	Emission function of unit i
$H_i(g_{i\tau})$	Emission function of unit i
I_c	Cell current (A)
$I_{i\tau}$	Commitment state (1 or 0) of unit i at hour τ
I_L	Light generated current (A)
I_o	Diode saturation current (A)
J	Segmentation number
J_i	Segmentation number can be served by i units
k	Boltzmann Constant (1.3854×10^{-23} JK^{-1})
K_o	Temperature difference between PV cell and ambient at wind speed zero (^oK)
K_1	The temperature gradient of PV cell (^oK sec/m)
K_i	Maximum fuel allocation for unit i

l_i	Latest period to begin maintenance of generator unit i
l'_k	Latest period to begin maintenance of line k
$m_g(i, \tau)$	Energy to supply load bid slope of generator i at hour τ
M_k	Number of lines that need maintenance in right-of- way k
MSR_i	Maximum sustain ramp rate of unit i , (MW/min)
n_c	Number of contingencies
n_τ	Number of hours in a week
N_{kt}	Number of candidate lines to be operated in right-of-way k in week t; in the vector form is $\boldsymbol{N}$; in hour is $N_{k\tau}$
$\overline{N}_k$	Maximum number of lines allowed in right-of-way k
N_p	Number of PV cells in parallel
Npv	Number of power states for PV
N_s	Number of PV cells in series
NB	Number of buses in the power system
NE	Number of segmentation in EEF
$NR_{i\tau}$	Non-Spinning reserve from generator i at hour τ
$NR_o(\tau)$	Total GENCO self-supplied non-spinning reserve at hour τ
$\overline{NR(\tau)}$	Maximum limit for non-spinning reserve at hour τ
p_i	Availability of unit i
$\overline{\boldsymbol{P}}^k(i)$	PV daily average power in day i with sd=k
$\overline{P}(G_t)$	Maximum power at radiation G_t
$P_b(\tau)$	Charge/discharge power to the battery at hour τ
$P_D(\tau)$	Total real power demand of the system at hour τ
$P_{Dj}(\tau)$	Load demand at bus j at hour τ
P_f	Fix charged for battery equalization
$P_{pv}(\tau)$	Instantaneous PV power output at hour τ
$P_s(\tau)$	PV spillage power at hour τ
$P_u(\tau)$	Total intermittent power injected to the network at hour τ
$\overline{P}_b$	Maximum charge/discharge power of battery

$\overline{P}_u$	Maximum intermittent power injected to the network
PC	STS production cost
$P\phi_\tau$	Power injection equivalent of phase shifter at hour τ
q	Charge of an electron ($1.6021\text{x}10^{-19}$As)
q_i	Forced outage rate of unit i, SCUC: quick start capability of unit i
$\overline{Q}_i$	Maximum reactive power of unit i
$\underline{Q}_i$	Minimum reactive power of unit i
$Q_D(\tau)$	Total reactive power demand at hour τ
$\overline{Qb}_i$	Upper limit reactive power injection at bus i
$\underline{Qb}_i$	Lower limit of reactive power injection at bus i
$\overline{Q}_T$	Upper limit equivalent reactive power injection of tap changer
$\underline{Q}_T$	Lower limit equivalent reactive power injection of tap changer
r_{it}	Real power interruption at bus i in week t; in vector form is $\boldsymbol{r}$; in hour is $r_{i\tau}$
$r_{si}(\tau)$	Contribution of unit i to spinning reserve at hour τ
$\overline{R}(\tau)$	Maximum limit for spinning reserve at hour τ
$R_{i\tau}$	Spinning reserve from generator i at hour τ
$R_0(\tau)$	System operating reserve requirement at hour τ PBUC: Total GENCO's self-supplied spinning reserve at hour τ
R_s	Series resistance (Ω)
$R_s(\tau)$	System spinning reserve requirement at hour τ
R_{sh}	Shunt resistance (Ω)
s	State index
$\overline{s}$	(maximum state) - (state index)
$s_{i\tau}$	Reactive power interruption or ancillary services at bus i in hour τ, in vector form is $\boldsymbol{s}$
sd	Number of storage days
$\boldsymbol{S}$	Node-branch incidence matrix
$S_{i\tau}$	Start up cost of unit i at hour τ

t	Time
$t_{oi}(\tau)$	Contribution of unit i to operating reserve at hour τ
T	PV cell temperature (°K) or tap changer position
$\overline{T}$	Upper limit of tap changer position
$\underline{T}$	Lower limit of tap changer position
T_{amb}	Ambient temperature (°K)
$T_i^{\text{on/off}}(\tau)$	Minimum up/down time of unit i
Tp	Power period
UR_i	Ramp up rate of unit i, (MW/hr)
V	System voltage in vector form
$V_b(\tau)$	Battery voltage at hour τ
V_c	PV cell voltage (volt)
W_{mt}	Maximum fuel m allocation in week t
W_{mj}	Maximum fuel m allocation in month j
W_{mv}	Maximum fuel m allocation in year v
x_{it}	Unit maintenance status, 0 if unit is off-line for maintenance; in vector form is $\boldsymbol{x}$
$X_{i(t-1)}^{off}$	Number of weeks that unit i has been on maintenance at week t-1
y_{kt}	Line maintenance status in week t, 0 if M_k line in right-of-way k is off-line for maintenance; in hour is $y_{k\tau}$
$Y_{k(t-1)}^{off}$	Number of weeks that lines k have been on maintenance at week t-1 at hour is $Y_{k(\tau-1)}^{off}$
$Z_{i(\tau)}^{on/off}$	Time in which unit i has been on/off at hour τ
α_i	Integrates labor starting up cost and equipment maintenance cost of unit i
β_i	Starting up cost of unit i from cold conditions
χ_t	Cost per MWh of energy purchased from outside sources at week t
$\chi_{i\tau}$	Power to be purchased to supply the GENCO bilateral contracts at generation bus i and at hour τ
δ_τ	Phase shifter angle at hour τ
ΔP	Power increment

ε	Acceptable level of expected energy not served
ϕ_{imt}	MBtu of fuel from the *m*-th contract (fuel *m*) allocated to unit *i* during week *t*
γ_k	Susceptance of branch *k*; in vector form is γ
$\eta(\tau)$	Instantaneous efficiency of PV
$\eta_b(\tau)$	Efficiency of battery at hour τ
φ	Probabilistic vector that defines the state of the system
ν_w	Win speed (m/sec)
θ	Voltage angle associated to node *i*; in vector form is θ
Θ_i	Time constant that characterizes unit *i* cooling speed
$\rho_g(i, \tau)$	Bid for energy to supply load by generator *i* at hour τ
$\rho_o(i, \tau)$	Marginal price of generator *i* at hour τ
$\rho_s(i, \tau)$	Bid for spinning reserve by generator *i* at hour τ
$\rho_{ns}(i, \tau)$	Bid for non-spinning reserve by generator *i* at hour τ
$\rho_\chi(i, \tau)$	Bid for power to be purchased at generation bus *i* and at hour τ

PREFACE

The overall goal of this book is to introduce algorithms for improving the economic posture of a utility company in a restructured power system by promoting cost-effective maintenance schedules.

Today, cutting operations and maintenance (O&M) costs and preserving service reliability[1] are among the top priorities for managers of utility companies. Preventive maintenance is perhaps the single largest controllable cost of a utility[2] operation. It is perceived that a careful planning and a good coordination among self-interested entities in a restructured power system are essential to achieving an optimal trade-off between the cost of maintenance and the service reliability. Traditional maintenance programs in vertically[3] integrated utilities relied heavily on time-directed maintenance and manufacturer recommendations. This book offers a logical alternative to traditional electric utility maintenance practices and a basis for maintenance decisions.

The book is organized as follows. Chapter I reviews various issues related to the power system operation and presents the role of restructuring in maintenance scheduling. In Chapter II, fundamental topics related to linear and nonlinear systems are reviewed. The duality in linear programming is discussed and integer programming is reviewed. Benders decomposition, Lagrangian relaxation, and Dantzig-Wolfe decomposition are presented. Several examples are given to demonstrate the applications of different methods. The formulation of reactive power optimization is discussed which will be used again in Chapter VII.

In Chapter III, the formulation of long-term generating unit maintenance scheduling is given which includes a GENCO's maintenance cost as the objective function, and numerous network and maintenance constraints such as flow constraints, availability of crews and other resources, as well as maintenance windows. Here, since we deal with independent entities (such as GENCOs and the ISO), Benders decomposition is used to solve the generating unit maintenance formulation. The proposed technique decomposes the original problem into a master problem, which is a relaxation of the original problem, and several independent sub-problems. The results for a simple system are analyzed and the application of the proposed method to IEEE-RTS is discussed. The complex fuel dispatch and emission constraints are included later and the network is modeled as a probabilistic problem to include the effect of generation and transmission outages. Further results illustrate the proposed comprehensive generation maintenance scheduling.

Chapter IV describes the details of short-term generation scheduling. First, a security constrained unit commitment (SCUC) package with real and reactive power

---

[1] Reliability of a system is interpreted as satisfying two major functions: adequacy and security, where an adequate amount of capacity resources should be available to meet the peak demand (adequacy) and the system should be able to withstand changes or contingencies on a daily and hourly basis (security).

[2] Utility industry in the United States is a $200 billion per year business.

[3] Vertical integration is an arrangement where the same company owns all the different aspects of making, selling and delivering a product or a service.

constraints is presented. Two different systems are used to demonstrate the efficiency of the SCUC package. Next, the formulation of price-based unit commitment (PBUC) is proposed for trading energy and ancillary services in the restructured market. The objective in PBUC is to maximize the GENCO's revenues. The short-term generation scheduling formulation is used in Chapter V.

Chapter V discusses the coordination between short-term and long-term generation scheduling problems in a GENCO. This chapter links Chapters III and IV. The results show that some of the solutions presented in Chapter III will encounter short-term scheduling violations if we do not consider the short-term constraints in the long-term maintenance scheduling of generating units.

Chapter VI discusses the problem of long-term transmission maintenance scheduling in a restructured system. The chapter presents the formulation of long-term transmission maintenance scheduling with probabilistic constraints. The coordination of TRANSCOs and the ISO for satisfying network flow and reliability constraints is discussed. A large set of constraints is included in the formulation and a step by step calculation is discussed for a simple power system. The application of the proposed method to IEEE-RTS is also presented.

Chapter VII presents the coordination between short-term and long-term transmission maintenance scheduling problems. The formulation includes cost-revenue tradeoffs and constraints that impact transmission line maintenance scheduling. The long-term maintenance of transmission lines is scheduled and the resulting maintenance windows and other variables are passed on to the short-term scheduling problem, which will determine the appropriate times within the given windows to perform maintenance. The method provides a dynamic schedule for calculating short-term maintenance periods. The IEEE 118-bus system network with 186 lines is tested and results are presented

Chapter VIII discusses the coordination of GENCOs' and TRANSCOs' maintenance schedules through the ISO. The solution will satisfy the objectives of these self-interested entities, as well as reliability and network flow constraints. The results point out that without the ISO's coordination, some of the objectives and/or constraints will not be satisfied in maintenance scheduling.

Chapter IX discusses the impact of unconventional sources of energy (i.e., photovoltaic with battery storage) on short-term generation scheduling. The chapter presents a method to simulate the production cost of GENCOs using a probabilistic approach, and a pilot project in the Eastern Islands of Indonesia is evaluated. Several constraints including the battery capacity, minimum up/down time and ramp rates for thermal units as well as photovoltaic (PV) capacity are considered in the proposed model. A new approach is also considered for incorporating PV-battery in thermal unit commitment. By incorporating battery storage, we can reduce load following requirements in the PV-utility grid. Furthermore, we can keep out peaking generators during peak hours by utilizing PV-battery. The formulation developed in this study is very flexible and can be applied to other renewable energy sources with intermittent natures. A case study composed of 26 thermal units and a PV-battery plant is presented. The short-term generating unit scheduling results presented in Chapter IX would enhance those of Chapter IV.

ACKNOWLEDGMENTS

We would like to take this opportunity to acknowledge the efforts of several people who guided us in writing this book.

We were inspired by Professor M.A. Pai (University of Illinois, Urbana) to write this book. His support of our ideas and his encouragement throughout this process are greatly appreciated. Many of the earlier thoughts for writing this book were based on the authors' conversations with Siemens engineers in Minneapolis. We also acknowledge the review of our book proposal and very constructive comments provided by Dr. Jay Giri (ALSTOM-ESCA Corporation), James Waight (Siemens Corporation), Professor Muwaffaq Alomoush (Yarmouk University of Jordan) and two of our senior research associates at Illinois Institute of Technology (Dr. Hatim Yamin and Dr. Yaoyu Wang). The last three individuals were very instrumental in editing several of the chapters in this book, and their unconditional support is genuinely appreciated. In addition, we acknowledge the editorial support of Mr. Alex Greene of Kluwer.

The first author has been fortunate to work with many engineers and educators throughout his twenty year career in power engineering. However, the two most significant individuals who have inspired him greatly to learn and research more about electric power systems are Dr. John Endrenyi (Ontario Hydro) and Dr. Bruce Wollenberg (University of Minnesota). These two individuals continue to inspire him with technical ideas as time progresses.

The second author would like to thank his colleagues at ABB, Drs. Arthur Cohen, Show Chang and Vladimir Brandwajn, for invaluable discussions on electric power restructuring paradigm.

This book could not have been completed without the moral support of our respective families. Their understanding and sacrifice are gratefully recognized.

M. Shahidehpour, Chicago, Illinois
M. Marwali, Santa Clara, California

CHAPTER I

INTRODUCTION

The restructuring of electric power industry has resulted in market-based approaches for unbundling a multitude of services provided by self-interested entities such as power generators (GENCO), transmission providers (TRANSCO), distribution companies (DISCO) and a host of others. As these entities move toward restructured market-based operation, decision-making tools must be prepared to evaluate the impact of competition. Market competition has created additional challenges for power companies to choose proper operational planning procedures for scheduling their equipment maintenance. The choice must take into account coordination between long-term and short-term maintenance schedules, as well as complex cost-revenue tradeoffs involved in evaluating the impact of maintenance outages on restructured utilities.

It is perceived that the restructuring will provide a major forum for the application of decomposition techniques to coordinate the optimization of various objectives among the self-interested entities. Earlier in the 1960-1970, many of the decomposition techniques were motivated by inability to solve large-scale centralized problems with the available computing power of that time. The dramatic improvement in computing technology since then allowed power engineers to solve very large problems much more easily. Consequently, interest in decomposition techniques dropped dramatically. However, now there is an increasingly important class of optimization problems in restructured power systems for which decomposition techniques are becoming most relevant.

In principle, one may consider optimizing a system of independent entities by constructing a large-scale mathematical program and solving it centrally (e.g., through the ISO) using currently available computing power and solution techniques. In practice, however, this is often impossible. In order to solve a problem centrally, one needs the complete information on local objective functions and constraints. As these entities are separated geographically and functionally, this information may be unattainable or prohibitively expensive to retrieve. More importantly, independent entities may be unwilling to share or report their private information as it is not incentive compatible to do so; i.e., these entities may have an incentive to misrepresent their true preferences.

In order to optimize certain objectives in restructured power systems, one must turn to the coordination aspects of decomposition. Specifically, with limited information one must coordinate entities to reach an optimal solution. Here, one identifies resources over which the entities compete and then charges each entity based on its resource usage. The goal will be to coordinate the entities by finding equilibrium resource prices while satisfying local and system constraints.

This chapter will discuss issues related to vertically integrated utilities and review the essence of restructured power systems. It will also discuss issues related to the operation and control of power systems and will specifically elaborate on maintenance scheduling and coordination in restructured utilities.

1.1 VERTICALLY INTEGRATED UTILITIES

Electric utilities operated as monopolies for many decades when they exercised exclusive control over the sale of electricity within their service territories. The control encompassed three components of the electricity business: generation, transmission and distribution. In addition, utilities controlled metering and billing services. During the era of monopoly, various regulations, laws and regulatory organizations were formed to oversee the activities of electric utilities in order to protect the interests of both consumers and the utilities. The supervision included review and approval of tariffs (or rates) charged for electric energy in addition to establishing and enforcing rules for providing utility-related services.

In a vertically integrated monopoly, utilities created regional dispatch power pools to coordinate the operation and planning of generation and transmission among their members in order to improve efficiencies and reliability and reduce costs. This was done through the least-cost mix of generation and transmission capacity, coordinating the maintenance of generating units, sharing operating reserves, and thus lowering the cost to end-use customers. During this cost-of-service regulatory operation, system operators preserved system reliability by ensuring that generation and load were matched on moment-to-moment basis (load following service) as load values could not be predicted with certainty. Vertically integrated utilities operated their own system, performed the economic dispatch of generation and managed trades with other utilities.

Investor-Owned Utilities (IOUs), which are also known as private utilities, are for-profit corporations that have provided retail services to the majority of customers in a vertically integrated monopoly. IOUs are owned by stockholders and operated by private individuals to gain shareholders' profit. Shares of IOUs are traded on public stock markets.

Under the structure of vertically integrated utilities shown in Figure 1.1, each utility:

- Controls and owns all or most of generation, transmission and distribution facilities within its geographical region.
- Exercises a monopoly on selling electric power to customers within its geographic region.
- Is obligated to sell and meet the needs of electric power to its customers.
- Is either publicly owned and not operated for profit, or has rates (prices) that are set by regulatory organizations.

In this structure, utilities are interconnected, through tie-lines, to form the power grid. In this grid, high-voltage transmission lines transmit power from remote areas, where large central station generating facilities are located, to load centers. Other functions of transmission lines include facilitating transfers of energy from one control area to another to share reserve capacities, take advantage of load diversity, increase system reliability, and transfer power at times of emergency or for economic reasons.

The Public Utility Regulatory Policy Act (PURPA)[1] of 1978, which was a federal legislation, required utilities to buy electric power, at an avoided cost[2] rate, from Independent Power Producers (IPPs) with private Qualifying Facilities (QFs)[3] as shown in Figure 1.2. This act was the beginning of restructuring in electric utilities.

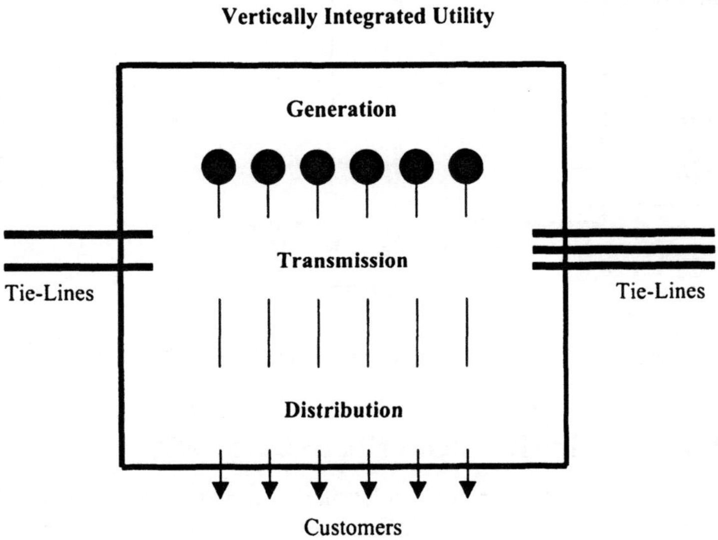

Figure 1.1 Power System Structure under Vertically Integrated Utilities

[1] PURPA is an act that was issued by the Congress in 1978. The act intended to encourage cleaner, more energy-efficient power production. It has created a new class of non-utility generators called qualifying facilities (QFs), that must meet certain ownership, size, and efficiency criteria established by FERC. A QF can sell its power to a utility at a price that is no higher than the cost the utility would have incurred had it supplied the power itself or obtained from another source (the avoided cost). PURPA promotes energy efficiency and increased use of alternative energy sources by encouraging companies to build co-generation facilities and renewable energy projects using wind power, solar energy, geothermal energy, hydropower, biomass, and waste fuels.

[2] Avoided Costs are the costs that a utility avoids by purchasing power from other power prividers (such as independent power producers) rather than generating the power itself or building new power plants. The Public Service Commission calculates avoided costs for each utility which are based on the cost of producing electricity by independent power producers. There are two parts to an avoided cost calculation: the avoided capacity cost of constructing new power plants and the avoided energy cost of fuel and operating and maintaining a utility's power plants.

[3] Qualifying Facility (QF) is an individual or corporation who owns and/or operates a small generation facility that qualifies under PURPA to supply electric energy and generating capacity that regulated utilities must purchase at a price that is no higher than the cost that the utility would have had to pay to produce the electricity itself or the cost it would have had to incur to purchase the power from another source (avoided cost). QF generators include those power producers that use renewable and alternative energy resources such as hydro, wind, solar, geothermal energy, biomass, municipal solid waste, or landfill gas fuel to generate power. Other QF power producers include co-generators.

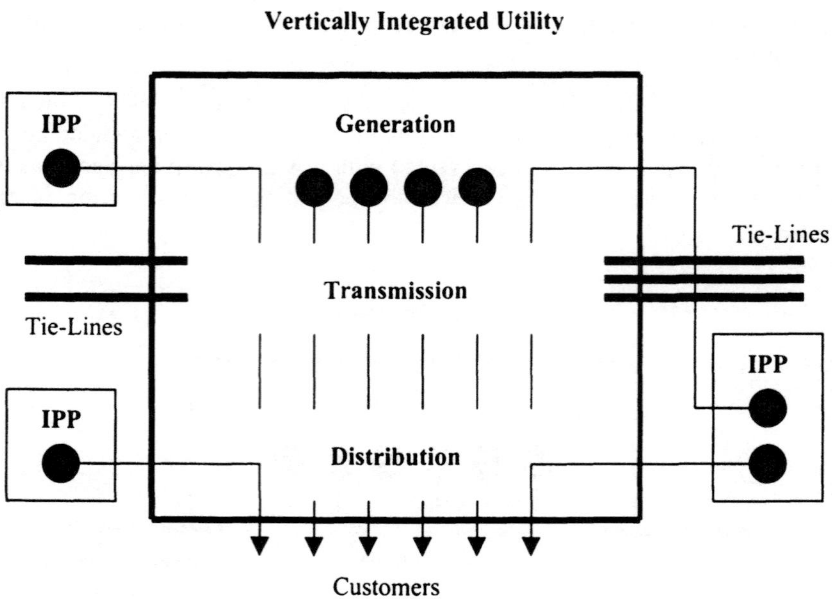

Figure 1.2 Power System Structure under the PURPA of 1978

1.2 RESTRUCTURED POWER INDUSTRY

The main aim of restructuring is to let market forces drive the price of electric supply and reduce the net cost through increased competition. Restructuring creates an open market environment by allowing the power supply to function competitively, as well as allowing consumers to choose the supplier of electric energy.

During the last few years, after the National Energy Policy Act[4] (EPAct, 1992), competition in the generation sector has been increased due to growth in the number of power suppliers; at the same time, open access to transmission system was more limited. Limitations came as vertically integrated utilities favored their own generation when transmission was congested, and limitation arose by preventing other utilities or suppliers full access to transmission systems. In addition, traditional power pools which controlled the access to regional transmission systems made it difficult for non-members to use pool members' transmission facilities by establishing complex operating rules and financial arrangements. Also, restrictive membership and governance of pools were practiced occasionally in a way that large utilities prevented changes in policies and rules of the pool which led to closing pool membership to outsiders.

---

[4] Energy Policy Act (EPAct) is a federal legislation established in 1992 that caused fundamental changes in the electric utility industry by promoting competition in wholesale electricity markets. The Act made it easier for non-utility producers to participate in the electric generation market. EPAct authorized FERC to mandate utilities to provide access to their transmission lines to other utilities, non-utility producers, and other participants in the wholesale electricity market.

Unfair industry practices generally impacted the growth of a competitive generation market and were motivating forces for the Federal Electric Regulatory Commission (FERC) to order transmission owners to provide other parties open access to transmission grids. However, in some instances transmission customers were not entitled to services provided by transmission owners. In this regard, FERC had to review requests on a case-by-case basis which might have taken more time and led to improper timing of permission to use the grid.

FERC Orders No. 888 and 889 were issued to bring restructuring to the electric power industry, which implied wholesale electricity competition, accelerating competition, bringing lower prices and creating more choices to consumers. The FERC Order 888 requested transmission owners to provide a comparable service to other customers who did not own any transmission facilities. Transmission owners were required by this order to treat external energy trades over their transmission facilities according to the same transmission tariffs that they applied to their owns. After this order, each transmission owner filed a *pro forma* tariff to implement the Order 888 by specifying terms and conditions of transmission services applicable to all customers. The FERC Order 889 required all market participants to obtain pertinent transmission information from the Open Access Same-time Information System (OASIS).

Restructuring necessitates the decomposition of three components of electric industry, and the separation of transmission ownership from transmission control as the best application of *pro forma* tariff. In addition, restructuring requires that metering and billing services, which were performed by vertically integrated utilities, be performed by entities on a competitive basis. A competitive generation market and retail direct access also necessitated independent operational control of the transmission grid. However, independent operation of the grid would not be guaranteed without an independent entity, the so-called Independent System Operator (ISO). The ISO is independent of individual market participants such as transmission owners, generators, distribution companies and end-users. The basic purpose of this entity is to ensure fair and non-discriminatory access to transmission services and ancillary services, and maintain real-time operation of the system and its grid reliability.

Restructuring is a very complex process and differs from place to place for many reasons, but in general, supply is the only part that is currently being restructured, while transmission and distribution functions will continue as regulated functions. These changes to the electric industry are initially required by the National Energy Policy Act and then by the FERC Notice of Proposed Rulemaking (NOPR) (which is also known as Mega-NOPR). A restructured power system is shown in Figure 1.3.

In this structure, generation companies (GENCOs) will be separately owned and compete to sell energy to customers, and may no longer be controlled by the same entities that control the transmission system. Transmission companies (TRANSCOs) will move power from place to place over high-voltage lines. Distribution companies (DISCOs) will move power at the retail level and may aggregate retail loads.

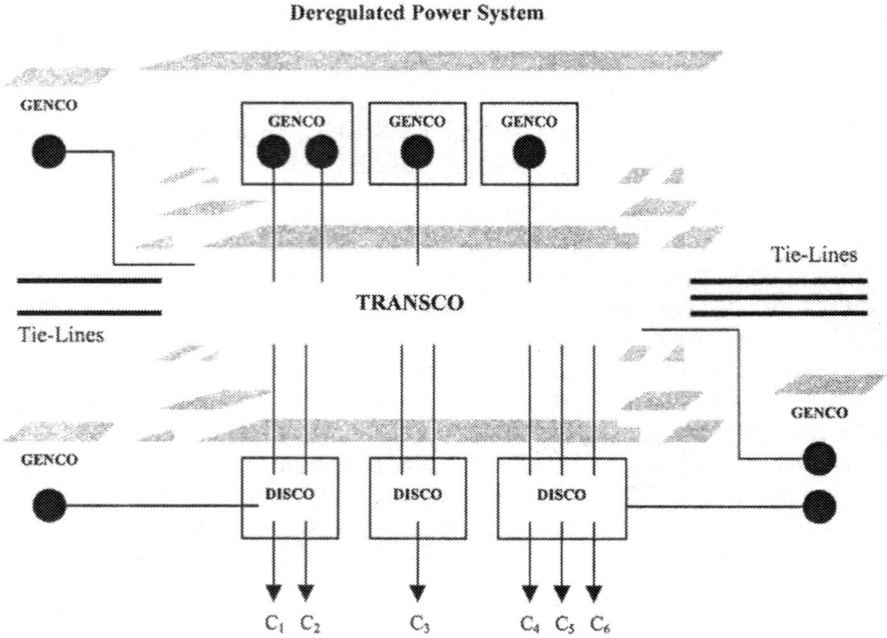

Figure 1.3 Power System Structure under the EPAct and the FERC Mega-NOPR

In general, new technical and business entities in a competitive electric marketplace are grouped into separate types of companies, which are GENCOs, TRANSCOs, DISCOs, Retail companies (RETAILCOs), Aggregators, Marketers, and Brokers. These entities, depicted in Figure 1.4, are discussed in the following sections of this chapter.

1.2.1 ISO. The ISO is a neutral operator responsible for maintaining instantaneous balance of the system. The ISO performs its function by controlling the dispatch of flexible plants and gives orders to adjust or curtail loads to ensure that loads match available generating resources in the system. The ISO is a voluntarily formed entity, which is developed on a regional basis due to differences among regions in operating and structural characteristics of the transmission grid. Although the ISO's responsibilities differ among restructuring models, in general, it is created to guarantee a comparable and non-discriminatory access by power suppliers and users to regional electric transmission systems.

The ISO is governed independent of any participants with commercial interests in the system operation. The ISO is granted the operational control of transmission grid components, administers a system-wide transmission tariffs, maintains short-term system reliability, coordinates maintenance scheduling, and has a role in coordinating long-term planning. An ISO should provide, to all eligible users, open access to the transmission system and all services under its control in a non-discriminatory manner.

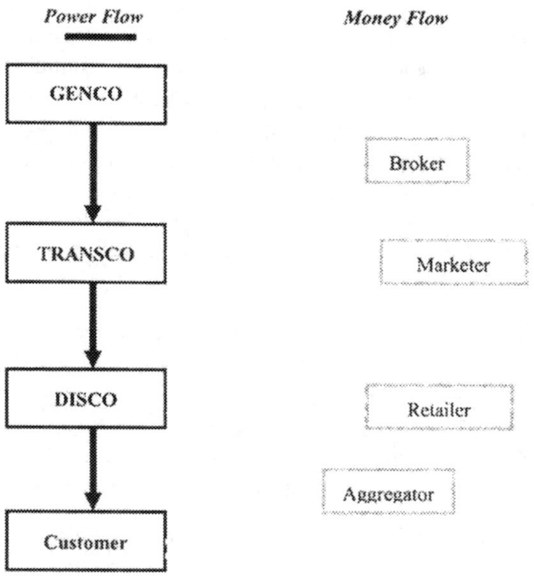

Figure 1.4 Relationships between Participants

The ISO is primarily responsible for ensuring the reliability of grid operations in a fashion that would comply with applicable standards set by NERC and the regional reliability council. The ISO has control over the operation of interconnected transmission facilities within its region, and has mechanisms to coordinate its function with neighboring control areas. It is the responsibility of the ISO to ensure an efficient trading among participants and create accurate price signals. The ISO has the ability to call for an increase or decrease in generation and to curtail loads for maintaining the system security.

A main task for the ISO is the management of transmission congestion including the collection and distribution of congestion revenues. During the process of mitigating transmission constraints, the ISO is responsible to ensure that proper economic signals are sent to all parties, to encourage an efficient use of and motivate investments in resources capable of alleviating constraints. Under emergency conditions, the ISO will have the authority to commit and dispatch some or all system resources or components.

The primary objective of the ISO is not to dispatch or re-dispatch generation, but to match the electricity supply with demand. The ISO should control generation to the extent necessary to maintain reliability, optimize transmission efficiency, and maintain stability of the power system. To maintain system integrity, the ISO is responsible for acquiring resources necessary to remove transmission violations, balance the system in second-to-second manner and maintain the system frequency at acceptable level to retain stability.

To comply with the FERC Order 889, the ISO is responsible for maintaining an electronic information network, i.e., OASIS, through which the current information on transmission systems is made available to all transmission system users such as transmission providers and transmission customers. Some capacity services are required to maintain the transmission system reliability, which should be contracted by the ISO with market participants in case they are not provided. These services include operating reserves, reactive power and frequency response capability to maintain the system frequency, capacity to cover losses and balancing energy. To make these services available, the ISO contracts with service providers so that the services are available under the ISO's request. Usually, the ISO chooses successful providers based on a least-cost bid basis.

In some market structures, the ISO and the power exchange (PX) are separate entities. While in other structures, the PX function is within the same organization and under the control of the ISO, and the ISO is responsible for dispatching all generators and setting the price of energy at each hour based on the highest price bid in the market. Most of the restructured models would include the PX and scheduling coordinators (SCs). Not all of these entities are present in all restructured models.

The PX may be merged with the ISO in some models, and the bilateral market represented by SCs may or may not exist in those models. Each market structure is characterized by authorities and activities delegated to the ISO. The PJM ISO and the National Grid Company (NGC) in the UK are examples of an ISO with a wide range of authorities and centralized control. The California ISO falls somewhere in the middle of the spectrum of restructuring where the ISO has no jurisdiction over forward energy markets and very limited control over actual generating unit scheduling. ERCOT and MAPP ISO's are examples of limited range of authorities for the ISO.

1.2.2 GENCOs. A GENCO is a regulated or non-regulated entity (depending upon the industry structure) that operates and maintains existing generating plants. GENCOs are formed once the generation of electric power is segregated from the existing utilities. A GENCO may own generating plants or interact on behalf of plant owners with the short-term market (power exchange, power pool, or spot market). GENCOs have the opportunity to sell electricity to entities with whom they have negotiated sales contracts. GENCOs may also opt to sell electricity to the PX from which large customers such as DISCOs and aggregators may purchase electricity to meet their needs. In addition to real power, GENCOs may trade reactive power and operating reserves. GENCOs are not affiliated with the ISO or TRANSCOs. A GENCO may offer electric power at several locations that will ultimately be delivered through TRANSCOs and DISCOs to customers.

GENCOs include IPPs, QFs, Exempt Wholesale Generators (EWGs)[5] created under EPAct, foreign utilities, and others. Generating assets include power

[5] An Exempt Wholesale Generator (EWG) is an independent power producer that generates power and sells it in the wholesale market at market-based rates and is exempt from restrictions normally imposed

producing facilities and power purchase contracts. Different from a vertically integrated structure, GENCOs will not be price-regulated. In addition, GENCOs are required to not discriminate against other market participants (such as DISCOs or RETAILCOs), fix prices or use bilateral contracts to exercise market power. GENCOs may be entitled to funds collected for the stranded power costs recovery.

GENCOs will communicate generating unit outages for maintenance to the ISO within a certain time (usually declared by the ISO) prior to the start of the outage. The ISO then informs the GENCOs of all approved outages. The role of GENCOs in committing generating units will be discussed later.

1.2.3 TRANSCOs. The transmission system is the most crucial element in electricity markets. The secure and efficient operation of the transmission system is the key to the efficiency in these markets.

A TRANSCO transmits electricity using a high-voltage, bulk transport system from GENCOs to DISCOs for delivery to customers. It is composed of an integrated network that is shared by all participants and radial connections that join generating units and large customers to the network. The use of TRANSCO assets will be under the control of the regional ISO, although the ownership is continued to be held by original owners in the vertically integrated structure. TRANSCOs are regulated to provide non-discriminatory connections and comparable service for cost recovery.

A TRANSCO has the role of building, owning, maintaining, and operating the transmission system in a certain geographical region to provide services for maintaining the overall reliability of the electrical system. TRANSCOs provide the wholesale transmission of electricity, offer open access, and have no common ownership or affiliation with other market participants (such as GENCOs and RETAILCOs). Authorities at state and federal levels regulate TRANSCOs which recover their investment and operating costs of transmission facilities using access charges (which are usually paid by every user within the area/region), transmission usage charges (based on line flows contributed by each user), and congestion revenues collected by the ISO.

The ISO handles the operation and scheduling of TRANSCOs' facilities. Transmission maintenance and expansion is coordinated between TRANSCOs and the ISO. TRANSCOs communicate with the ISO the list of equipment outages, or any changes to the scheduled outages, within a certain time (usually declared by the ISO) prior to the start of the outage. The ISO then informs TRANSCOs of all approved outages. Later, we will discuss the issue of transmission maintenance further when we talk about the ISO's role in coordination.

1.2.4 DISCOs. A DISCO is an entity that distributes the electricity, through its facilities, to customers in a certain geographical region. A DISCO is a regulated (by state regulatory agencies) electric utility that constructs and maintains distribution wires connecting the transmission grid to final customers.

---

on electricity providers by the Public Utility Holding Company Act (PUHCA). This type of company was created by the Energy Policy Act of 1992 in order to increase level of competition in wholesale electricity generation. FERC is responsible for determining EWG status.

DISCOs will continue to function in much the same fashion as distribution companies did, but with a greater service quality monitoring. A DISCO is charged with building and operating its electric system to maintain a certain degree of reliability and availability. In addition, a DISCO has the responsibility of responding to distribution network outages and power quality concerns. To ensure the flow of power, DISCOs coordinate their functions with TRANSCOs and the ISO to provide distribution services and perhaps bundled service to customers in their service territories. DISCOs are required to deliver the power to customers on a non-discriminatory basis at tariffed distribution service rates. DISCOs are responsible for maintenance and voltage support as well as ancillary services including the coordination with the ISO, and may perform metering, billing and collection services. DISCOs may also carry out other services such as aggregating customers, purchasing power supply and transmission services for customers, reimbursing suppliers, and offering other regulated or non-regulated energy services to retail customers. The wires and customer service functions provided by a DISCO could be split between two totally separate entities.

1.2.5 OASIS. The FERC Order 889 has mandated the establishment of OASIS using the Internet infrastructure. Each TRANSCO is required to establish an OASIS node alone or with other TRANSCOs. A TRANSCO may access an OASIS node using a web browser and request a firm/non-firm transmission reservation. A TRANSCO is also required to provide frequently on the OASIS the updated information on available transmission capability (ATC). ATC is a measure of the transfer capability remaining in the physical transmission network for further commercial activity over and above already committed uses.

OASIS supports the posting of ATCs as well as the offering of transmission and ancillary services with their associated prices and terms. OASIS allows users to reserve capacity on the transmission system, purchase ancillary services, re-sell transmission services to others, and buy ancillary services from third party suppliers. Services offered by OASIS differ from one market structure to another.

FERC and several state commissions have approved regional tariffs, market structures, and business processes that are standard within a particular region but are not uniform across the country. Under some emerging regional tariffs, either the transmission reservation provided by OASIS would not be required in the current sense or a different process may be used for reserving transmission capacity. Regional markets appear to becoming more diverse from each other rather than more similar.

1.2.6 RETAILCOs. A RETAILCO is a newly created entity in this competitive industry that obtains legal approval to sell retail electricity. A RETAILCO takes title to the available electric power and re-sells it in the retail customer market. A retailer buys electric power and other services necessary to provide electricity to its customers and may combine electricity products and services in various packages for sale. A retailer may deal indirectly with end-use customers through aggregators.

A RETAILCO has no organizational or financial affiliation with a DISCO. It will arrange for distribution of electricity by DISCOs and will be responsible for developing product offerings to customers and markets. A given RETAILCO will be able to work with many DISCOs and a DISCO may have many RETAILCOs doing business in its service area. Any type of entity can acquire the role of RETAILCO after meeting certain standards. RETAILCOs are required to not discriminate against customers or DISCOs. Retail consumers will have the opportunity to enter bilateral contracts for power with RETAILCOs. A GENCO may become a RETAILCO for this purpose or customers may form a RETAILCO. RETAILCOs may create an aggregation of services that would allow purchasing wholesale power and breaking it down into retail power sales, which may be perceived as providing least-cost services to customers.

1.2.7 Aggregator. An aggregator is an entity or a firm that combines customers into a buying group for buying large blocks of electric power and other services with a cheaper price. The aggregator may act as an agent (broker) between this group of customers and a retailer. When an aggregator purchases power and re-sells it to customers, it acts as a retailer and should initially qualify as a retailer.

1.2.8 Marketer. A marketer is an entity or a firm that buys and re-sells electric power but does not own generating facilities. A marketer takes title, and is approved by FERC, to market electric energy services. A marketer performs as a wholesaler and may acquire required transmission services. An entity may handle both marketing and retailing functions.

1.2.9 Broker. A broker of electric energy services is an entity or firm that acts as a middleman in a marketplace in which those services are priced, purchased, and traded. A broker does not take tittle on available transactions, and does not generate, purchase or sell electric energy but facilitates transactions between buyers and sellers. If a broker is interested in acquiring a title on electric energy transactions, then it should be classified as a generator or a marketer. A broker may act as an agent between a GENCO, or an aggregation of generating companies, and marketers.

1.2.10 Customer. A customer is the end-user of electricity with certain facilities connected to the distribution system, in the case of small customers, and connected to transmission system, in the case of bulk customers. In a vertically integrated structure, a user obtains electric energy services from a utility that has legal rights to provide those services in the service territory where the customer is located. In a restructured system, customers are no longer obligated to purchase any services from their local utility company. Customers will have direct access to generators or contracts with other providers of power, and choose packages of services (such as the level of reliability) with the best overall value that meets customers' needs. For instance, customers may choose providers that would render the option of shifting customer loads to off-peak hours with lower rates.

1.3 OPERATION AND MAINTENANCE IN A RESTRUCTURED POWER SYSTEM

Operation and maintenance costs are rising as increasing traffics strain the capabilities of power systems to maintain reliability. Electric utilities spend millions of dollars per year on maintenance to keep their facilities operational in the most efficient and secure state. Maintenance practices currently rely on time or some other counters such as number of operations and/or visual inspections. The existing maintenance approaches may be overly conservative and could result in maintenance being scheduled when none is required or deferred when it is critical. New maintenance practices are needed in restructured power systems that would allow managers to take proper steps to ensure reliability while controlling and, even lowering, costs.

Reliability and economy are among the most important considerations in restructured power systems, and maintenance scheduling of facilities has found a prominent position in the operation of power systems. Unnecessary equipment outages may incur economic losses and pose a serious danger to power systems and its security. An optimized maintenance schedule can improve system reliability, reduce system operation costs and introduce savings in capital investment for new facilities.

In a vertically integrated electric utility, it was the responsibility of the utility to make maintenance schedules for a variety of facilities. So, generating unit maintenance scheduling, transmission line maintenance scheduling as well as the coordination of schedules were done centrally. The exclusive advantage of this centralized processing was that the solution could optimize the reliability and operation cost of the entire system owned by the utility.

1.4 MAINTENANCE SCHEDULING IN A RESTRUCTURED POWER SYSTEM

Maintenance scheduling is a complicated discrete stochastic non-linear optimization problem, especially when independent entities such as GENCOs and TRANSCOs and their self-interested objectives are taken into account. In a broad sense, there are two kinds of facility maintenance in power systems: generating unit maintenance and transmission line maintenance. Theoretically, maintenance of generation and transmission may be studied independently. However, system constraints such as network flows limit and energy demand and reliability requirements will closely tie the two solutions to encourage practical optimization and feasible solutions for the two problems.

Maintenance scheduling establishes a preventive outage schedule within a specific time horizon, e.g. one year, for a number of on-line facilities. Generally, the solution process will divide the time horizon into small time intervals, like weeks, and then determine when each facility outage should begin in order to optimize a given objective.

The generating unit maintenance scheduling problem was first proposed when engineers tried to optimize the operational scheduling of a large power system about three decades ago. The transmission line maintenance scheduling problem has a much shorter history and was originally included as a constraint in the solution of generating unit maintenance. However, with additional complexities introduced in modern large-scale power systems, transmission line maintenance scheduling has become a separate subject. This is true especially in a restructured power system where generation units and transmission lines belong to totally different entities of the power market.

The functional separation of transmission and generation creates operational and scheduling problems related to maintenance. Additional bottlenecks arise in transmission maintenance as additional generation commitment and dispatch may be required to relieve congestion. Similar effects may arise when generating units are taken off-line for maintenance. Hence, maintenance schedules must be coordinated through an independent entity (i.e., the ISO) to assure reliable and economical service. The ISO in this environment will resolve conflicts between GENCOs and TRANSCOs regarding maintenance schedules, because competitive interests of GENCOs may not always agree with transmission maintenance schedules of TRANSCOs.

The ISO should cancel or reschedule planned transmission and generation outages if those outages impact the system severely. Transmission outages may impact the market due to major changes in transfer capabilities. To approve, cancel or reschedule planned outages, the ISO should develop an outage coordination protocol to cover both generator and transmission outage plans that would maintain system reliability and security, and consider economic market impacts on TRANSCOs' and GENCOs' assets. The ISO should also identify whether or not planned outages scheduling by various entities is intended for gaming in order to gain unfair market advantage or to exercise market power by pushing prices over competitive levels.

1.4.1 ISO's Function in Maintenance Scheduling. In its policy for system coordination[6] (Policy 4, Maintenance Coordination Section: Generator and Transmission Outages), NERC states that "Scheduled generator and transmission outages that may affect the reliability of interconnected operations shall be planned and coordinated among affected systems and control areas." The ISO would collect all transmission and generator planned outage requests from market participants, i.e., TRANSCOs and GENCOs. The ISO would then review all submissions of planned outages based on operating reliability criteria that conform with NERC reliability requirements, types of outage (forced, planned and unplanned) and the time/date of request for maintenance. The ISO would then decide whether to permit, deny, or adjust planned outage schedules to preserve the system reliability. The ISO may also coordinate planned transmission and generator outage schedules with neighboring systems and control areas to ensure that additional reliability criteria required by NERC would not be violated.

---

[6]NERC Operating Manual, http://www.nerc.com/~oc/opmanp2.html

Intense competition in power markets would result in a more complicated facility maintenance scheduling and create additional pressure on GENCOs and TRANSCOs to make up an optimal maintenance schedule for their facilities. For example, emission control requirements will impact off-line generating unit maintenance and on-line fuel allocation. With rapid developments in the restructuring of power systems, it is imperative to develop efficient decision–making tools for GENCOs and TRANSCOs to explore the most appropriate maintenance schedules in a competitive situation.

1.5 FORMULATION OF THE MAINTENANCE SCHEDULING PROBLEM

1.5.1 Optimization Objectives. The reliability and economic objectives often conflict with the operation of power systems, and are very difficult to be integrated in one mathematical formulation. In most cases, it seems more acceptable to consider the cost as an optimization objective and the reliability as a constraint.

An optimal generating unit maintenance schedule in a GENCO should satisfy the following requirements:

- Reduce the operation cost of the GENCO
- Maintain the reliability of the system
- Extend the life span of generating units
- Save investment costs for the installation of new generating units

The operation cost of a GENCO can be further divided into the energy production cost and the maintenance cost. The energy production cost is a function of the amount of fuel burnt by units. The maintenance cost can be curtailed when the planned outage schedule is partitioned and adjusted according to the availability of resources and manpower. [Dop75] concluded that the generating unit maintenance cost is almost a convex function, and for each generating unit there is an ideal schedule for maintenance. [Zur75] pointed out that the difference between the most expensive and the least expensive generating unit maintenance schedules was about 0.08% of the expected production cost, which is rather substantial.

GENCOs' objectives are to extend the life span of units through proper maintenance and make profit by trading energy. TRANSCOs' objectives are to maintain transmission lines in a reliable condition and make profit by wheeling energy. The ISO's responsibility is to guarantee the system reliability and leave out GENCOs' or TRANSCOs' profit as a consideration. Therefore, it is very difficult to seek a common optimization objective for facility maintenance scheduling in a restructured environment. Moreover, because the objectives and constraints of these entities are totally different, their respective formulation and solution methods of the problem will be quite different.

In essence, the objective-driven characteristic of the problem in a restructured environment has determined that the solution methodology will be quite different from that in the traditionally regulated power systems. Therefore, efficient analytical methods and effective decision making tools should be provided to each relevant participant, and any optimization solution for the facility maintenance scheduling problem must encompass GENCOs, TRANSCOs and the ISO with their respective objectives.

1.5.2 Problem Constraints. In vertically integrated utilities, because the facilities maintenance scheduling problem is considered centrally, many constraints are taken into account simultaneously. Listed below are some of the constraints for maintenance scheduling:

- Number of maintenance crews
- Maintenance period during which a facility can be maintained
- Maintenance interval: the length of time required for maintenance
- Sequence precedence: some facilities should be arranged for maintenance in advance of the others
- Repetitive maintenance: some facilities may need maintenance more often in a given period
- Load demand
- Rated capacity of the facility
- Geographical constraint: maximum number of facilities that can be in maintenance in each region
- Forced outage rate and reliability of the facility
- System reliability

In a restructured power system, facility maintenance is scheduled by the facility owner and approved by the ISO. Hence, the corresponding owner in the power market will be primarily concerned with many of the constraints. For example, GENCOs will consider constraints related to generation units, such as the energy price especially the predictable seasonal energy price when they make generating unit maintenance schedule. So once a GENCO has determined the proper schedule for its facility maintenance, it will forward the schedule to the ISO which will check network flow conditions based on the proposed generating unit maintenance schedule. The transmission network conditions and availability will be supplied by TRANSCOs. This topic will be discussed further in Chapter 3.

In order to schedule a practical solution for maintenance, the viability of generation and transmission facility maintenance outages ought to be examined by the ISO in an integrated facility maintenance scheduling. These topics and solutions are discussed in detail throughout this book.

1.5.3 Solution Methods. Generating unit maintenance scheduling has a long history as engineers have studied this subject for about three decades. In a vertically integrated power system, the generating unit maintenance scheduling problem was formulated as a very complicated discrete stochastic non-linear optimization problem, in which the transmission line maintenance scheduling problem was generally represented as a constraint. The most common solution for this problem was to first construct an implicit function of independent variables and then perform a complex simulation to compute the objective function for a large number of potential schedules. Various optimization methods were applied to solve the problem, which could be classified into the following three large categories:

- Heuristic Methods: which provide the most primitive solution based on trial-and error principles.
- Artificial Intelligence (AI): which include expert system, simulated annealing, fuzzy theory, neural network, evolutionary optimization including evolutionary programming, evolutionary strategy and genetic algorithm, simulated evolution, Tabu search and various combinations of AI methods.
- Mathematical Programming: which include integer programming, mixed-integer linear programming, decomposition, branch-and-bound, dynamic programming and various combinations of mathematical programming methods.

Although in the past decades, several methods were proposed for the solution of unit maintenance scheduling in vertically integrated utilities, there was no consensus on the most appropriate approach to this problem. Earlier, much emphasis was given to heuristic methods because of their flexibility and simplicity. However, the heuristic methods could not meet the multi-objective requirements of the problem and could not guarantee a feasible solution.

Most AI techniques have the capability of dealing with multi-objective requirements. Since an inference engine must be designed according to the particular characteristics of a designated problem, it is difficult to generalize the expert system approach. The membership functions in fuzzy sets are to be configured according to the specific requirements of the designated power system, hence fuzzy sets are generally used as an auxiliary tool in maintenance optimization methods. However, the literature indicates that, of all the possible intelligence techniques, genetic algorithms are the most suitable AI technique for maintenance scheduling. There is no doubt in our minds that mathematical programming methods provide more reliable and versatile solutions to maintenance scheduling, and branch-and-bound and dynamic programming seem more suitable than others in a restructured system.

This book considers that the solution of the maintenance scheduling problem in a restructured power system is composed of two separate steps. In step one, a facility owner schedules the maintenance. In step two, the schedule is submitted to the ISO for testing its viability to satisfy system constraints. The book considers a variety of cases and constraints which involve GENCOs, TRANSCOs and the ISO. Since these entities are independent in restructured systems, the book considers mixed-integer programming and decomposition techniques to coordinate and optimize maintenance scheduling. The problem, in general, is split into one master problem and several sub-problems, which are solved iteratively until a proper solution is obtained. The simulation results show that the decomposition-coordination mechanism is a very practicable methodology for the facility maintenance scheduling problem in restructured power systems.

It is to be noted that this book discusses general mathematical approaches for solving a very complicated problem and it is expected that, with rapid developments in the restructuring of power industry, additional topics and issues on maintenance will arise later that will require further studies in the near future.

CHAPTER II

MATHEMATICAL REVIEW

This chapter reviews some of the fundamental topics related to linear and nonlinear systems optimization. Decomposition will be applied later to independent entities in power systems, which are represented by the ISO, GENCOs and TRANSCOs. Section 2.1 discusses the duality in linear programming. Section 2.2 reviews integer programming. Section 2.3 presents Benders decomposition. Section 2.4 reviews the Lagrangian relaxation method. Dantzig-Wolfe decomposition is presented in Section 2.5. An application of Dantzig-Wolfe decomposition to power loss minimization is presented in Section 2.6.

2.1 DUALITY IN LINEAR PROGRAMMING

Corresponding to any linear program, called the primal problem, there is another one called its dual. These two problems relate to one another with interesting applications. Let us consider the following two linear programming (LP) problems:

LP I

$$\text{Minimize} \quad z = \sum_{n=1}^{N} c_n x_n \tag{2.1}$$

$$\text{St.} \quad \sum_{n=1}^{N} a_{mn} x_n \geq b_m \qquad m=1,\ldots,M \tag{2.2}$$

$$x_n \geq 0 \qquad n=1,\ldots,N \tag{2.3}$$

LP II

$$\text{Maximize} \quad y = \sum_{m=1}^{M} b_m \pi_m \tag{2.4}$$

$$\text{St.} \quad \sum_{m=1}^{M} a_{nm} \pi_m \leq c_n \qquad n=1,\ldots,N \tag{2.5}$$

$$\pi_m \geq 0 \qquad m=1,\ldots,M \tag{2.6}$$

If x_n is given as the primal variable of LP I, then π_m is the dual variable of LP I. In LP, a dual variable is also called a simplex multiplier. We make a one-to-one correspondence between the mth dual variable π_m and the mth primal constraint (with the right-hand side b_i) and between the nth primal variable and the nth dual constraint (with the right-hand side c_i). Table 2.1 gives the primal-dual correspondence in which the dual of the dual will be the primal. π_m is a "shadow price" for item m and represents the increment in the objective value due to a small increase in item m. Note that π_m may be either positive or negative.

Table 2.1 Primal-Dual Correspondence

Primal quantity	**Corresponding dual quantity**
Objective $c_n \; x_n \rightarrow$ min	Objective $b_m \; \pi_m \rightarrow$ max
Variable $x_n \geq 0$	Constraint $a_{nm} \; \pi_m \leq c_n$, inequality
Variable x_n unconstrained in sign	Constraint $a_{nm} \; \pi_m = c_n$, equality
Constraint $a_{mn} \; x_n \geq b_m$, inequality	Variable $\pi_m \geq 0$
Constraint $a_{mn} \; x_n = b_m$, equality	Variable π_m unconstrained in sign
Coefficient matrix A	Coefficient matrix A^T
Right-hand side b_m	Right-hand side c_n
Cost coefficient c_n	Cost coefficient b_m

Properties of dual linear problems:

1. The primal problem has a finite optimal-feasible solution if and only if the dual has a finite optimal-feasible solution and, in such a case, optimal values will be the same.

2. When one problem has no feasible solution, then the other either has no feasible solution or has an unbounded optimal-feasible solution.

Example 2.1

Minimize $z = 50x_1 + 40x_2$

St.
$$3x_1 + 2x_2 \geq 35$$
$$5x_1 + 6x_2 \geq 60$$
$$2x_1 + 3x_2 \geq 30$$
$$x_1 , \; x_2 \geq 0$$

The solution is: $x_1 = 9 ; x_2 = 4 ; z = 610$.

The dual to the preceding problem is then:

Maximize $y = 35\pi_1 + 60\pi_2 + 30\pi_3$

St.
$$3\pi_1 + 5\pi_2 + 2\pi_3 \leq 50$$
$$2\pi_1 + 6\pi_2 + 3\pi_3 \leq 40$$
$$\pi_1 , \pi_2 , \pi_3 \geq 0$$

The solution is: $\pi_1 = 14, \pi_2 = 0 , \pi_3 = 4 , y = 610$.

In the above example, $\pi_2 = 0$ indicates that the second constraint of the primal problem will not bind the objective value. On the other hand, $\pi_1 = 14$ and $\pi_3 = 4$ indicate that the objective value will gain 18 units by relaxing the bounds on the first and the third constraints of primal problem by one unit. In other words, if we set the bounds of the first and the third primal constraints to 36 and 31 instead of 35 and 30 respectively, the objective will gain 18.

2.2 INTEGER PROGRAMMING

Branch-and-bound is an intriguing technique with extensive applications to large integer and mixed-integer problems. Consider the following mixed integer-program:

$$\text{Minimize } f(y) = \sum_{j=1}^{N} c_j y_j$$

$$\text{St.} \quad \begin{aligned} &\sum_{j=1}^{N} a_{ij} y_j \le b_i \quad i = 1,2,\dots,M \\ &U_j \ge y_j \ge L_j \quad j = 1,2,\dots,N \\ &y_k \ge 0 \qquad k = 1,2,\dots,\overline{N} \\ &y_l \text{ integer} \quad l = \overline{N}+1, \overline{N}+2,\dots,N \end{aligned} \tag{2.7}$$

If $\overline{N}$ is zero, we have an integer linear program; if all variables are restricted to zero-one, we have a binary integer program. All sets of problems can be attacked using branch-and-bound procedures.

The underlying concept of branch-and-bound is the observation that at any value of y_j, one can always state that $[y_j]+1 \ge y_j \ge [y_j]$ where $[y_j]$ is the largest integer less than (or equal to) y_j. Suppose that we obtain $y_j^* = 4.3$ by the simplex method. Since the solution is supposed to be integer, we establish two mutually exclusive and exhaustive sets: those which satisfy $U_j \ge x_j \ge 5.0$ and those which satisfy $4.0 \ge x_j \ge L_j$. Using this observation, we create two new problems from our original problems, one that adds $U_j \ge x_j \ge 5.0$ to the original problem constraints and one that add $4.0 \ge x_j \ge L_j$. That is, we *branch* to new problems. Note further that this process of creating new problems rapidly creates a *tree* of possible alternatives.

Each feasible solution will represent a lower bound on the optimal solution. It should be clear that in the process of branching and possibly creating bounds on the optimal solution, feasible solutions satisfying all constraints might be found. In the event that no such solution can be found, there will exist no feasible solution.

Example 2.2

Maximize $f(y) = 9y_1 + 6y_2 + 5y_3$

St. $$2y_1 + 3y_2 + 7y_3 \le \frac{35}{2}$$

$$4y_1 \quad + 9y_3 \le 15$$

$$y_1 \quad \text{non-negative integer}$$

Step1: Solve the problem as an ordinary linear program with the following solution:

$$y_1 = 15/4, \quad y_2 = 10/3, \quad y_3 = 0, \quad f(y) = 215/4.$$

Step 2: This solution does not satisfy the integer restrictions, so we divide the original problem into the following two mutually exclusive and exhaustive subproblems:

(a) *Maximize* $f(y) = 9y_1 + 6y_2 + 5y_3$

St. $$2y_1 + 3y_2 + 7y_3 \le \frac{35}{2}$$

$$4y_1 \quad + 9y_3 \le 15$$

$$y_1 \quad \ge 4 \text{ (integer)}$$

$$y_2, y_3 \ge 0$$

(b) *Maximize* $f(y) = 9y_1 + 6y_2 + 5y_3$

St. $$2y_1 + 3y_2 + 7y_3 \le \frac{35}{2}$$

$$4y_1 \quad + 9y_3 \le 15$$

$$0 \le y_1 \quad \le 3 \text{ (integer)}$$

$$y_2, y_3 \ge 0$$

It is clear that problem (a) has no feasible solution. The solution to problem (b) is given by: $y_1 = 3, y_2 = 23/6, y_3 = 0$ and $f(y) = 50$. Since this satisfies all problem restrictions, it is a feasible and optimal solution. To further illustrate the branch-and - bound procedure, consider the following similar problem:

Maximize $f(y) = 9y_1 + 6y_2 + 5y_3$

St. $$2y_1 + 3y_2 + 7y_3 \le \frac{35}{2}$$

$$4y_1 \quad + 9y_3 \le 15$$

$$y_1, y_2, y_3 \ge 0 \quad \text{and all are integers}$$

Step 1: First we obtain the following linear programming solution:

$$y_1 = 15/4\ ,\ y_2 = 10/3\ ,\ y_3 = 0\ ,\ f(y) = 215/4 \quad \textit{(not all integers)}$$

Step 2: Choosing variable y_1, we create two new problems, (a) and (b). One has no feasible solution and one generates the following solution:

$$y_1 = 3,\ y_2 = 23/6,\ \ y_3 = 0,\ \ f(y) = 50 \quad \textit{(not all integers)}$$

Step 3: The problem in (b) of step 2 can be divided into two more problems:

(c) *Maximize* $f(y) = 9y_1 + 6y_2 + 5y_3$

$$\begin{aligned}
\textit{St.} \qquad & 2y_1 + 3y_2 + 7y_3 \le \frac{35}{2} \\
& 4y_1 \qquad + 9y_3 \le 15 \\
& 0 \le y_1 \le 3 \ \textit{(integer)} \\
& y_2 \ge 4 \\
& y_3 \ge 0
\end{aligned}$$

(d) *Maximize* $f(y) = 9y_1 + 6y_2 + 5y_3$

$$\begin{aligned}
\textit{St.} \qquad & 2y_1 + 3y_2 + 7y_3 \le \frac{35}{2} \\
& 4y_1 \qquad + 9y_3 \le 15 \\
& 0 \le y_1 \le 3 \ \textit{(integer)} \\
& 0 \le y_2 \le 3 \\
& y_3 \ge 0
\end{aligned}$$

Problem (c) has no feasible solution, while the solution to problem (d) is given by: $y_1 = 3,\ y_2 = 3,\ y_3 = 0,\ \ f(y) = 45$. Since all variables at this solution are integer, the procedure is terminated.

2.3 BENDERS DECOMPOSITION

The Benders decomposition algorithm is developed for exploiting mixed-integer programs. Benders decomposition has many applications in a restructured power system, which is represented by a series of independent entities.

In applying the Benders decomposition, the original problem is formulated in the form of (2.8). Based on the linear programming duality theory, the original problem will be decomposed into a master problem and a subproblem. The process of solution of the master problem begins with only a few (or no) constraints. The subproblem is used as a test to see if this solution satisfies the remaining constraints. If so, the solution is optimal, since the objective has been minimized over all constraints. If not, a constraint which is most unsatisfied will be added to the master problem and the master problem will be re-solved with more constraints. The final solution based on the Benders algorithm may require iterations between the master problem and the subproblem. Since variable y is usually integer, the master-program is an integer problem. The subproblem is the linear programming problem which tests the optimality of a solution to the master problem and, if necessary, provides the most unsatisfied constraint, i.e., the deepest cut, to the master problem.

2.3.1 Formulation of Benders Decomposition.

The mixed-integer problem has the following form:

$$\left.\begin{array}{ll}\text{Minimize} & \boldsymbol{c}^{\mathrm{T}}\boldsymbol{x}+f(\boldsymbol{y})\\ \text{St.} & \boldsymbol{A}\boldsymbol{x}+\boldsymbol{F}(\boldsymbol{y})\geq \boldsymbol{b}\\ & \boldsymbol{x}\geq 0,\quad \boldsymbol{y}\in S\end{array}\right\}\quad \text{Problem P1} \tag{2.8}$$

where,

$\boldsymbol{A}$: $m\times n$ matrix,
$\boldsymbol{x}$, $\boldsymbol{c}$: n vectors,
$\boldsymbol{y}$: p integer vector,
f : scalar-valued function of $\boldsymbol{y}$,
$\boldsymbol{F}$: m vector whose components are functions of $\boldsymbol{y}$,
$\boldsymbol{b}$: m vector,
S : an arbitrary subset of E^p with integral-valued components

Since $\boldsymbol{x}$ is continuous and $\boldsymbol{y}$ is integer, (P1) is a mixed-integer problem. If values of y are fixed, (P1) is linear in $\boldsymbol{x}$. Hence, (2.8) could be written as:

$$\underset{y\in R}{miniminze}\left\{f(\boldsymbol{y})+min\left\{\boldsymbol{c}^T\boldsymbol{x}\mid \boldsymbol{Ax}\geq \boldsymbol{b}-\boldsymbol{F(y)},\boldsymbol{x}\geq 0\right\}\right\} \tag{2.9}$$

where

$$R=\left\{\boldsymbol{y}\mid \text{ there exists } \boldsymbol{x}\geq 0 \text{ such that } \boldsymbol{Ax}\geq \boldsymbol{b}-\boldsymbol{F}(\boldsymbol{y}),\ \boldsymbol{y}\in S\right\} \tag{2.10}$$

Based on the duality theory, the set R in (2.10) may be rewritten

$$R=\left\{\boldsymbol{y}\ \middle|\ (\boldsymbol{b}-\boldsymbol{F}(\boldsymbol{y}))^T\ \boldsymbol{u}_i^r\quad i=1,\ldots\ldots,n_r,\ \ \boldsymbol{y}\in S\right\} \tag{2.11}$$

$\boldsymbol{u}_i^r$ is the extreme point vector which belongs to cone $C=\left\{\boldsymbol{u}\middle| A^T\boldsymbol{u}\leq 0,\boldsymbol{u}\geq 0\right\}$. n_r is number of extreme points of cone C. The inner minimization in (2.9) can be rewritten as follows:

Primal

$$\begin{array}{ll}Miniminze & \boldsymbol{c}^T\boldsymbol{x}\\ St. & \boldsymbol{Ax}\geq \boldsymbol{b}-\boldsymbol{F}(\boldsymbol{y})\\ & \boldsymbol{x}\geq 0\end{array} \tag{2.12}$$

Its dual problem is:

$$\begin{array}{ll}Maximize & (\boldsymbol{b}-\boldsymbol{F}(\boldsymbol{y}))^T\ \boldsymbol{u}\\ St. & A^T\boldsymbol{u}\leq \boldsymbol{c}\\ & \boldsymbol{u}\geq 0\end{array} \tag{2.13}$$

Substituting (2.13) into (2.9) yields a new form for P1:

$$\underset{y \in R}{Miniminze}\left\{ f(y) + max\left\{(b - F(y))^T u \,\middle|\, A^T u \le c, u \ge 0\right\}\right\} \tag{2.14}$$

Program (2.14) is equivalent to the following program (P2)

$$\left.\begin{array}{l} \text{Minimize } z \\ \text{St. } z \ge f(y) + (b - F(y))^{\mathrm{T}} u_i^p,\ i = 1,\ldots,n_p \\ \quad (b - F(y))^{\mathrm{T}} u_i^r \le 0 \quad i = 1,\ldots,n_r \\ \quad y \in S \end{array}\right\} \text{ Problem P2} \tag{2.15}$$

where u_i^p is an extreme point of:

$$P = \left\{ u \,\middle|\, A^{\mathrm{T}} u \le c, u \ge 0 \right\} \tag{2.16}$$

Problem (P2) in (2.15) is equivalent to problem (P1) in (2.8) with integer variable y. (P2) has one constraint for each extreme point which may be an enormous number even in a problem with moderate dimensions. However, only a small fraction of constraints will be binding at an optimal solution. Therefore, we begin with a few (or no) constraints and solve the resulting modified (P2), i.e., MP2 in (2.17). Subproblem (2.12) or (2.13) is used to see if this solution satisfies the remaining constraints. The modified MP2 problem is as follows:

$$\left.\begin{array}{l} \text{Minimize } \quad z \\ \text{St. } \quad z \ge f(y) + (b - F(y))^{\mathrm{T}} u_i^p,\ i \in I_1 \\ \quad (b - F(y))^{\mathrm{T}} u_i^r \le 0 \quad i \in I_2 \\ \quad y \in S \end{array}\right\} \text{ Problem MP2} \tag{2.17}$$

where I_1 and I_2 are proper subsets of integers $1,\ldots.n_p$ and $1,\ldots.n_r$ respectively. The resulting algorithm involves iterations between two problems. The first is the modified master problem to which unsatisfied constraints are successively added. The second are linear programming subproblems which test the optimality of a solution to the modified problem and, if necessary, provide a new constraint.

After solving (2.17), we obtain solution z^0, y^0. This solution is optimal for P2 if and only if:

$$(b - F(y^0))^{\mathrm{T}} u_i^p \le z^0 - f(y^0) \qquad i = 1,\ldots.n_p \tag{2.18}$$

$$(b - F(y^0))^{\mathrm{T}} u_i^r \le 0 \qquad i = 1,\ldots..n_r \tag{2.19}$$

It is easy to prove that (2.17) is the dual of (2.12) as:

$$\begin{aligned} \text{Maximize} \quad & (\boldsymbol{b}-\boldsymbol{F}(\boldsymbol{y}))^{\mathrm{T}} \boldsymbol{u} \\ \text{St.} \quad & \boldsymbol{A}^{\mathrm{T}} \boldsymbol{u} \leq \boldsymbol{c} \\ & \boldsymbol{u} \geq 0 \end{aligned} \tag{2.20}$$

So, if the solution of (2.20) is equal to $z^0 - f(y^0)$, the solution of (2.17) is the solution of the original problem. Otherwise, we add the new constraints (2.18) to MP2 and solve it again. If (2.20) has an unbounded solution, the simplex method locates an extreme ray and extreme point u such that the dual objective approaches infinite. Since, $(\boldsymbol{b}-\boldsymbol{F}(y^0))\boldsymbol{v}^0 \geq 0$ is not satisfied, y^0 will not satisfy $(\boldsymbol{b}-\boldsymbol{F}(y))\boldsymbol{v}^0 \leq 0$. Hence, we add this constraint to MP2 and solve it again.

2.3.2 Steps of the Algorithm.

1. Initiate the procedure by establishing MP2, where only a few (or no) constraints of the form (2.17) are included.

2. Solve MP2. If the solution of MP2 is infeasible, so will be P1 and P2. Otherwise, we obtain either a finite optimal solution or information that the solution is unbounded. If unbounded, set $z^0 = \infty$ in (2.17), let y^0 be an arbitrary element of S and go to step 3.

3. Solve the dual linear subprogram (2.13) [or the primal (2.12) if it is feasible]. If the dual is infeasible, then the original problem P1 will have an unbounded solution. If the dual is unbounded, go to step 6

4. If the optimal value in step 3 is $z^0 - f(y^0)$, the solution of P2 is (z^0, y^0). If the solution of (2.15) is $\boldsymbol{x}^0$, then the solution of P1 is $(\boldsymbol{x}^0, y^0)$.

5. If the optimal test in step 3 is not passed and the dual (2.16) has a finite optimal solution, $\boldsymbol{u}^0$, then, $z^0 < f(y^0) + (\boldsymbol{b}-\boldsymbol{F}(y^0))^{\mathrm{T}} \boldsymbol{u}^0$. So, the current solution of MP2 does not satisfy $z \geq (\boldsymbol{b}-\boldsymbol{F}(y))^{\mathrm{T}} \boldsymbol{u}^0 + f(y)$. Add this constraint to MP2 and return to step 2.

6. If the dual (2.13) has an unbounded solution, the simplex method locates an extreme ray and extreme point u such that the dual objective would approach infinite. Since $(\boldsymbol{b}-\boldsymbol{F}(y^0))\boldsymbol{v}^0 > 0$ is satisfied, y^0 would not satisfy $(\boldsymbol{b}-\boldsymbol{F}(y^0))\boldsymbol{v}^0 > 0$. Add this constraint to MP2 and return to step 2.

The flowchart for the Benders decomposition is as shown in Figure 2.1.

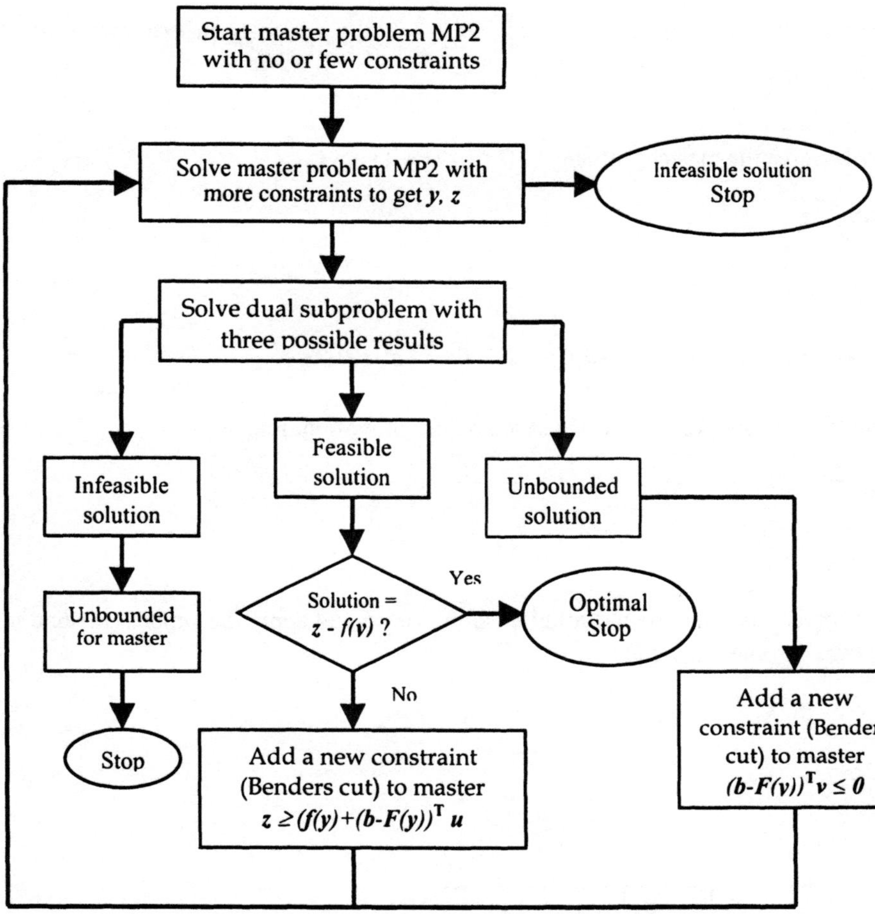

Figure 2.1 Benders Decomposition Flowchart

2.3.3 Example for Benders Decomposition.

Example 2.3

$$\min_{x\geq 0, y\geq 0} \quad x_1 + 3x_2 + y_1 + 4y_2$$

$$St. \quad -2x_1 - x_2 + y_1 - 2y_2 \geq 1$$

$$2x_1 + 2x_2 - y_1 + 3y_2 \geq 1$$

$$c^T = [1\ 3] \qquad f = [1\quad 4] \quad F = \begin{bmatrix} 1 & -2 \\ -1 & 3 \end{bmatrix} \quad A = \begin{bmatrix} -2 & -1 \\ 2 & 2 \end{bmatrix} \qquad b = \begin{bmatrix} b_1 \\ b_2 \end{bmatrix} = \begin{bmatrix} 1 \\ 1 \end{bmatrix}$$

Start the MP2 solution with $u^0 = \begin{bmatrix} u_1^0 & u_2^0 \end{bmatrix}^T = [0\ 0]^T$ which is an extreme point of cone P (2.16).

Step 1 Solve the master problem MP2 (2.17) with $u_i^p = \begin{bmatrix} 0 \\ 0 \end{bmatrix}$

$$\underset{x \geq 0, y \geq 0}{Min} \quad z$$

$$St. \quad z \geq y_1 + 4y_2$$

The optimal solution is: $y_1 = 0 \quad y_2 = 0 \quad z^0 = 0$.

Solve the dual subproblem (2.13) based on the above solution:

$$\underset{u \geq 0}{\text{Max}} \quad 2u_1 + u_2$$

$$\text{St.} \quad -2u_1 + 2u_2 \leq 1$$

$$-u_1 + 2u_2 \leq 3$$

This problem has an unbound solution. Go to step 6 and apply the simplex method to the primal problem (2.12) to get v_0:

$$\underset{u \geq 0}{Max} \quad x_1 + 3x_2$$

$$St. \quad -2x_1 - x_2 \geq 1$$

$$2x_1 + 2x_2 \geq 1$$

		x_1	S_4
z	-1.5	2.0	1.5
S_3	1.5	1.0	0.5
x_2	0.5	-1.0	-0.5

So, $v_0 = \begin{bmatrix} 1 \\ 0.5 \end{bmatrix}$ and the new Benders cut $(b - F\,y)^T v_0 \leq 0$ is $y_1 - y_2 \geq 3$.

Step 2 Solve the Benders master problem (2.17) with the above new Benders cut:

$$\underset{x \geq 0, y \geq 0}{Min} \quad z$$

$$St. \quad z \geq y_1 + 4y_2$$

$$y_1 - y_2 \geq 3$$

The optimal solution is $y_1 = 3,\ y_2 = 0,\ z^0 = 3$. Solve the dual problem (2.13):

$$\underset{u \geq 0}{Max} \; -2u_1 + 4u_2$$

$$St. \quad -2u_1 + 2u_2 \leq 1$$

$$-u_1 + 2u_2 \leq 3$$

$u^0 = [2.0 \;\; 2.5]$ $\;z_u^0 = 6$ is the feasible solution.

Since $z^0 - f(y^0) = 3 - [1 \;\; 4]\begin{bmatrix} 3 \\ 0 \end{bmatrix} = 0 \neq z_u^0 = 6$, the optimal test (2.18) is not satisfied. So the new constraint added to MP2 (2.17) is $z \geq 1.5y_1 - 0.5y_2 + 4.5$.

Step 3 Solve the Benders master problem (2.17) with the new constraints:

$$\underset{x \geq 0, y \geq 0}{Min} \; z$$

$$St. \quad z \geq y_1 + 4y_2$$

$$y_1 - y_2 \geq 3$$

$$z \geq 1.5y_1 - 0.5y_2 + 4.5$$

$[y_1 \;\; y_2] = [3 \;\; 0]$, $z^0 = 9$. We solve primal and dual problems (2.12), (2.13) to find $[x_1 \;\; x_2] = [0 \;\; 2]$, $z_u^0 = 6$. Because $z^0 - f(y^0) = 9 - [1 \;\; 4]\begin{bmatrix} 3 \\ 0 \end{bmatrix} = 6 = z_u^0 = 6$, the optimal test (2.18) is satisfied.

Hence, the final solution is: $[x_1 \;\; x_2] = [0 \;\; 2]$, $[y_1 \;\; y_2] = [3 \;\; 0]$, $z_{opt} = 9$.

2.4 LAGRANGIAN RELAXATION

2.4.1 Introduction. Lagrangian relaxation is based on the observation that many integer programming problems complicated by a set of constraints can be modeled as a relatively easy Lagrangian problem in which the complicating constraints are replaced with a penalty term in the objective function representing the violation of constraints and their dual variables. The Lagrangian problem provides an upper/lower bound (for a maximum/minimum problem) on the optimal value of the original problem. It can thus be used in place of a linear programming relaxation to provide bounds in a branch and bound algorithm.

In order to apply the Lagrangian relaxation method, the mathematical optimization problem should be written in the following form:

$$\begin{aligned} &Min \quad f(x) \\ &St. \quad g(x) = 0 \\ &\qquad h(x) \leq 0 \\ &\qquad x \geq 0 \end{aligned} \tag{2.21}$$

The constraints in (2.21) are partitioned into three sets so that the problem can be solved easily once the constraints $g(x)=0$, $h(x)\leq 0$ are removed. To create the Lagrangian problem, we define non-negative λ and μ., as Lagrange multipliers. The Lagrangian relaxation function is defined by relaxing the above constraints into the objective function except for the bound constraints. Each constraint that is relaxed into the objective function will get a separate Lagrange multiplier. The Lagrangian function is then defined as follows:

$$\Gamma(x,\lambda,\mu) = f(x) + \lambda g(x) + \mu h(x) \tag{2.22}$$

The mathematical optimization problem (2.21) is written in terms of the Lagrangian function as follows:

$$\begin{aligned} &Min \ \Gamma(x,\lambda,\mu) = f(x) + \lambda g(x) + \mu h(x) \\ &\text{St.} \quad x \geq 0 \end{aligned} \tag{2.23}$$

There are different techniques such as Kuhn-Tucker to solve the Lagrangian relaxation optimization problem (2.23). In this chapter, we apply the dual optimization technique to solve the Lagrangian relaxation problem.

2.4.2 Dual Optimization Technique.

We apply dual optimization to convex and non-convex optimization problems as discussed in the following sections.

Consider the following problem:

$$\begin{aligned} &Min \quad c^T x + f(y) \\ &St. \quad Ax + F(y) \geq b \\ &\qquad x \geq 0, y \in S \end{aligned} \tag{2.24}$$

By dualizing $Ax + F(y) \geq b$, we obtain the following Lagrangian function $\Gamma(x,\lambda)$:

$$\Gamma(x,\lambda) = Min_{x\geq 0}\, c^T x + f(y) + \lambda(b - Ax - F(y)) \tag{2.25}$$

If x does satisfy $Ax+F(y)\geq b$, then $\Gamma(x,\lambda)$ will be non-positive and, thus, for all $\lambda\geq 0$, we have:

$$\Gamma(\mathbf{x},\lambda) = Min_{x\geq 0}\, c^T x + f(y) + \lambda(b - Ax - F(y)) \tag{2.26}$$

The Lagrangian dual q^* is then defined by $q^* = \max_{\alpha \geq 0} q(\lambda)$.

$$q(\lambda) \leq Min_{x \geq 0}\, c^T x + f(y) + \lambda(b - Ax - F(y)) \tag{2.27}$$

Under certain conditions sufficient for strong duality,

$$q(\lambda) = Min_{x \geq 0}\, c^T x + f(y) + \lambda(b - Ax - F(y)) \tag{2.28}$$

which enables us to find the Lagrangian dual as:

$$q^* = Max_{\lambda \geq 0}\left(Min_{x \geq 0}\, c^T x + f(y) + \lambda(b - Ax - F(y))\right) \tag{2.29}$$

In the case of convex optimization, the dual problem will provide the solution for the primal problem. However, in case of non-convex optimization, the objective function solution of the dual problem at the optimal point will provide a lower bound to the objective function solution of the primal problem. Most of the power systems mathematical problems are non-convex.

The procedure that is used for setting up the multipliers is called multiplier-adjustment technique. This heuristic technique is used widely to solve non-differentiable problems. The technique, used in the dual problem, exploits special features of the dual problem in certain applications.

The following are the steps for solving the optimization problem based on the multiplier-adjustment technique:

1. Start with setting all Lagrange multipliers λ to zero.
2. Optimize on variables x to get J^*. If constraints are not satisfied then set J^* to a high number. Next, calculate q^*.
3. Check the relative duality gap $(J^* - q^*)/q^*$, and if its value is less than a specified tolerance then stop. Usually, the relative duality gap reaches zero in convex optimization problems.
4. Update Lagrange multipliers of violated constraints using a gradient adjustment: $\lambda^{new} = \lambda^{old} + \left[\frac{d}{d\lambda} q(\lambda)\right].k$. Where k is a problem dependent constant. It is advised to allow λ to be adjusted downward at a slower rate than upward.
5. Go to step 1.

Lagrangian relaxation is applied widely to unit commitment, which determines start up and shut down periods of thermal generating units. Also, it determines the power output of generating units to meet the demand so that the total operating cost is minimized while satisfying the operational constraints. In unit commitment, there are integer variables, I_i, restricted to 1 or 0. Having discrete variables in the optimization problem makes it a non-convex problem. Mathematically, unit commitment is formulated as a mixed-integer nonlinear optimization problem and Lagrangian relaxation is most suitable since the resulting relaxed primal problem can be decomposed into one subproblem for each generating unit. So, the solution of unit commitment is obtained by solving many small size subproblems. This technique would allow us to utilize parallel computing techniques with a remarkable CPU time.

The application of the dual optimization technique to a non-convex problem with discrete variables is explained by the following example.

Example 2.4

$$\begin{aligned} &Min\ \ I_1 f_1(x_1) + I_2 f_2(x_2) \\ &\text{St.}\ \ I_1 x_1 + I_2 x_2 = 170 \\ &\quad 100 \le x_1 \le 300 \\ &\quad 50 \le x_2 \le 200 \\ &\quad I_1, I_2 = 0 \text{ or } 1 \end{aligned}$$

where,

$$f_1(x_1) = 500.0 + 10.0x_1 + 0.002x_1^2$$

$$f_2(x_2) = 100.0 + 6.0x_2 + 0.005x_2^2$$

Solution

The Lagrangian relaxation function is as follows:

$$\Gamma(x_1, x_2, I_1, I_2, \lambda) = I_1 f_1(x_1) + I_2 f_2(x_2) + \lambda(170 - I_1 x_1 - I_2 x_2)$$

The dual problem is presented as $q^* = \max_{\lambda \ge 0} q(\lambda)$

where, $q(\lambda) = \min_{x_1, x_2, I_1, I_2} \Gamma(x_1, x_2, I_1, I_2, \lambda)$.

$$\text{Assume } k = \begin{cases} 0.01 & if\ \dfrac{dq}{d\lambda} > 0 \\ 0.002 & if\ \dfrac{dq}{d\lambda} < 0 \end{cases}$$

We set $J^* = 10{,}000.0$ if the constraint is violated and assume that the satisfactory relative duality gap is 0.8. We proceed with the following iterations:

Iteration 0.

Set initial Lagrange multipliers to zero, $\lambda = 0.0$. So, the Lagrangian relaxation function will be:

$$\Gamma = I_1 f_1(x_1) + I_2 f_2(x_2)$$

Here, there are two terms that are summed up in the Lagrangian function. We minimize the Lagrangian function by minimizing each term separately. For each term, the minimum will be zero if $I = 0$ and negative if $I = 1$.

$$\frac{d}{dx_1}[f_1(x_1)] = 0.0 \longrightarrow x_1 = -2500.$$

Since the lower bound is 100, then $x_1 = 100$.

$$[f_1(x_1)]_{at\ x_1=100} > 0.0 \longrightarrow I_1 = 0$$

$$\frac{d}{dx_2}[f_2(x_2)] = 0.0 \longrightarrow x_2 = -600$$

Since the lower bound is 50, then $x_2 = 50$.

$$[f_2(x_2)]_{at\ x_2=50} > 0.0 \longrightarrow I_2 = 0$$

$$q^* = 0 \quad and \quad J^* = 10{,}000.0$$

Then, the relative duality gap, $\frac{10{,}000.0 - 0.0}{0.0}$, is unacceptable and we proceed with additional iterations.

Iteration 1.

$$\lambda^1 = \lambda^0 + \left[\frac{d}{d\lambda}q(\lambda)\right].k = 0.0 + (0.01)(170) = 1.7$$

Now, the Lagrangian relaxation function is:

$$\Gamma = I_1[f_1(x_1) - \lambda x_1] + I_2[f_2(x_2) - \lambda x_2] + 170\lambda^1$$

The last term is fixed, so we minimize each part of Γ individually as follows:

$$\frac{d}{dx_1}[f_1(x_1) - \lambda x_1] = 0.0 \longrightarrow x_1 = -2075$$

Since the lower bound is 100, then $x_1 = 100$.

$$[f_1(x_1) - \lambda x_1]_{at\ x_1=100} > 0.0 \longrightarrow I_1 = 0$$

$$\frac{d}{dx_2}[f_2(x_2) - \lambda x_2] = 0.0 \longrightarrow x_2 = -430$$

Since the lower bound is 50, then $x_2 = 50$.

$$[f_2(x_2) - \lambda x_2]_{at\ x_2=50} > 0.0 \longrightarrow I_2 = 0$$

$$q^* = 289 \quad and \quad J^* = 10{,}000.0$$

The relative duality gap, $\dfrac{10{,}000.0 - 289.0}{289.0} = 33.6$, is higher than the specified value. So, we proceed with more iterations.

Iteration 2.

$$\lambda^2 = \lambda^1 + \left[\frac{d}{d\lambda} q(\lambda)\right].k = 1.7 + (0.01)(170) = 3.4$$

Now, the Lagrangian relaxation function will be:

$$\Gamma = I_1 [f_1(x_1) - \lambda x_1] + I_2 [f_2(x_2) - \lambda x_2] + 170\lambda^2$$

The last term is fixed, we minimize each part of Γ individually as follows:

$$\frac{d}{dx_1}\left[f_1(x_1) - \lambda x_1\right] = 0.0 \longrightarrow x_1 = -1650$$

Since the lower bound is 100, then $x_1 = 100$.

$$\left[f_1(x_1) - \lambda x_1\right]_{at\ x_1 = 100} > 0.0 \longrightarrow I_1 = 0$$

$$\frac{d}{dx_2}\left[f_2(x_2) - \lambda x_2\right] = 0.0 \longrightarrow x_2 = -260$$

But since the lower bound is 50, then $x_2 = 50$

$$\left[f_2(x_2) - \lambda x_2\right]_{at\ x_2 = 50} > 0.0 \longrightarrow I_2 = 0$$

$$q^* = 578 \quad and \quad J^* = 10{,}000.0$$

The relative duality gap, $\dfrac{10{,}000.0 - 578.0}{578.0} = 16.3$, is higher than the specified value. So, we proceed with more iterations. We repeat the same procedure for iterations 3 and 4 and no change occurs in discrete variable values. Also, the relative duality gap is still higher than the specified value.

Iteration 5.

$$\lambda^5 = \lambda^4 + \left[\frac{d}{d\lambda} q(\lambda)\right].k = 8.5$$

Now, the Lagrangian relaxation function will be:

$$\Gamma = I_1 [f_1(x_1) - \lambda x_1] + I_2 [f_2(x_2) - \lambda x_2] + 170\lambda^5$$

The last term is fixed, we minimize each part of Γ individually as follows:

$$\frac{d}{dx_1}\left[f_1(x_1)-\lambda x_1\right]=0.0 \longrightarrow x_1=-375$$

Since the lower bound is 100, then $x_1=100$.

$$\left[f_1(x_1)-\lambda x_1\right]_{at\ x_1=100}>0.0 \longrightarrow I_1=0$$

$$\frac{d}{dx_2}\left[f_2(x_2)-\lambda x_2\right]=0.0 \longrightarrow x_2=250$$

Since the upper bound is 200, then $x_2=200$.

$$\left[f_2(x_2)-\lambda x_2\right]_{at\ x_2=200}<0.0 \longrightarrow I_2=1$$

So, we substitute the calculated discrete variables in the primal problem and then calculate $x_2=170.0$ which satisfies the constraint. The calculated optimal values are: $q^*=1264.5 \quad and \quad J^*=500+10(170)+0.002(170)^2=2257.8$

Then, the relative duality gap is $\dfrac{2257.8-1264.5}{1264.5}=0.7855$ which is less than the specified value. So, we stop here.

We will discuss the application of Lagrangian relaxation to unit commitment in Chapter IV.

2.5 DANTZIG-WOLF DECOMPOSITION

Dantzig-Wolfe decomposition is an efficient procedure when applied to linear problems whose coefficient matrices have an angular structure, i.e., independent blocks linked by coupling equations. The solution is based on a technique called "column generation" which is used to generate columns that would enter basis by solving subproblems. Dantzig-Wolfe decomposition operates by forming an equivalent "master program" and a set of independent subproblems. The resulting algorithm involves iterations between subproblems, whose objective functions contain variable parameters, and the master program. Subproblems receive a set of parameters (simplex multipliers or prices) from the master program and then send their solutions to the master program, which combines these with previous solutions in an optimal way and computes new prices and send them back to the subproblems. The iteration proceeds until an optimality test is passed. The Dantzig-Wolfe algorithm is reviewed briefly here [Las70].

2.5.1 Formulation of Dantzig-Wolfe Decomposition.

The development of the Dantzig-Wolfe decomposition is based on the following premises:

1. Column generation: If the constraint coefficients matrix of a linear program has many rows or columns, then the Dantzig-Wolfe decomposition principle can be used to reduce the number of rows at the expense of more columns. This can be accomplished by using the pivot operation on the constraint coefficients matrix.
2. Theorem: A point is in a closed bounded convex polyhedron if and only if it can be written as a convex combination of extreme points of the polyhedron.

Let us consider a linear programming problem with the following p-block angular structure form:

$$\text{Minimize} \quad z = \boldsymbol{c}_1\boldsymbol{x}_1 + \boldsymbol{c}_2\boldsymbol{x}_2 + \ldots\ldots + \boldsymbol{c}_p\boldsymbol{x}_p \tag{2.30}$$

$$\begin{aligned} St. \quad & \boldsymbol{A}_1\boldsymbol{x}_1 + \boldsymbol{A}_2\boldsymbol{x}_2 + \ldots\ldots + \boldsymbol{A}_p\boldsymbol{x}_p = \boldsymbol{b}_0 \\ & \boldsymbol{B}_1\boldsymbol{x}_1 = \boldsymbol{b}_1 \\ & \quad \boldsymbol{B}_2\boldsymbol{x}_2 = \boldsymbol{b}_2 \\ & \quad \quad \ldots \\ & \quad \quad \quad \boldsymbol{B}_p\boldsymbol{x}_p = \boldsymbol{b}_p \\ & \boldsymbol{x}_1 \geq 0, \quad \boldsymbol{x}_2 \geq 0, \quad \ldots\ldots \quad \boldsymbol{x}_p \geq 0 \end{aligned} \tag{2.31}$$

where $\boldsymbol{c}_i$, $\boldsymbol{x}_i$ are subvectors, $\boldsymbol{A}_i$, $\boldsymbol{B}_i$ are submatrices and $i = 1,2,\ldots\ldots p$. The first row identifies the coupling constraints and the remaining rows represent the independent blocks.

Based on the above theorem, the equivalent master program is:

$$\text{Minimize} \quad \sum_{i,j} f_{ij}\lambda_{ij} \tag{2.32}$$

$$St. \quad \sum_{i,j} \boldsymbol{p}_{ij}\lambda_{ij} = \boldsymbol{b}_0 \tag{2.33}$$

$$\sum_{j} \lambda_{ij} = 1, \quad i = 1,\ldots,p, \quad \lambda_{ij} \geq 0 \tag{2.34}$$

where $f_{ij} = \boldsymbol{c}_i\boldsymbol{x}_i^j$

$\boldsymbol{p}_{ij} = \boldsymbol{A}_i\boldsymbol{x}_i^j$

$\boldsymbol{x}_i^j$ is an extreme point of convex polyhedron $S_i = \{\boldsymbol{x}_i | \boldsymbol{B}_i\boldsymbol{x}_i = \boldsymbol{b}_i, \boldsymbol{x}_i \geq 0\}$

The Dantzig-Wolfe decomposition principle solves the optimization problem at the subproblem level, but not at the master program level, where only a single pivot operation is performed.

The restricted master program is a more symmetric formulation with all columns dropped, except those in the current basis and the one that is about to enter. To avoid tabulating all the columns, the column generation technique is used to create columns to enter the basis.

Then, the restricted master program is:

$$\text{Minimize} \quad \sum_{i=1}^{p} \sum_{j(basic)} f_{ij}\lambda_{ij} + \sum_{i=1}^{p} f_i^* \lambda_i^* \tag{2.35}$$

$$\text{St.} \quad \sum_{i=1}^{p} \sum_{j(basic)} p_{ij}\lambda_{ij} + \sum_{i=1}^{p} p_i^* \lambda_i^* = b_0 \tag{2.36}$$

$$\sum_{j(basic)} \lambda_{ij} + \quad \lambda_i^* = 1 \tag{2.37}$$

$$\lambda_{ij} \geq 0 \qquad \lambda_i^* \geq 0 \tag{2.38}$$

where p_i^* and f_i^* are the columns and relative cost factors generated from the last subproblem solutions. λ_{ij} is the current basic variable and λ_i^* is the entering variable.

Hence, the p independent subproblems become:

$$\text{Minimize} \quad z_i = (c_i - \pi A_i) x_i \tag{2.39}$$

$$\text{St.} \quad B_i x_i = b_i \qquad x_i \geq 0 \tag{2.40}$$

where π is the simplex multiplier from the revised simplex method.

2.5.2 Steps for the Decomposition Algorithm.

Assume that an initial basic feasible solution for the master program is available, with a basis matrix B and simplex multipliers $(\pi, \pi_{01}, \pi_{02}, \ldots \pi_{0p})$.

Step 1. Using the simplex multipliers π, solve p independent subproblems, and obtain solutions $x_i(\pi)$ and optimal objective values z_i^0. The ith independent subproblem is:

$$\text{Minimize} \quad z_i = (c_i - \pi A_i) x_i \tag{2.41}$$

$$\text{St.} \quad B_i x_i = b_i \qquad x_i \geq 0 \tag{2.42}$$

Step 2. Compute the minimal subproblem objective value,

$$\bar{f}_{ij} = (c_i - \pi A_i) x_i^j - \pi_{0i} \tag{2.43}$$

and if $$\min_{1\le i\le p}\min_{j} \bar{f}_{ij} = \min_{1\le i\le p}(z_i^0 - \pi_{0i}) \ge 0 \tag{2.44}$$

then the current solution is optimal. The optimal solution for (2.30)-(2.31) is:

$$\boldsymbol{x}^0 = \sum_{i}\sum_{j(basic)} \lambda_{ij}\boldsymbol{x}_i^j \tag{2.45}$$

where $\boldsymbol{x}_i^j$ is the extreme point of convex polyhedron S_i corresponding to basic λ_{ij}.

Step 3. If $\min_i \min_j \bar{f}_{ij} < 0$, then form the column:

$$\boldsymbol{p} = \begin{pmatrix} A_i\boldsymbol{x}_i(\pi) \\ \boldsymbol{u}_i \end{pmatrix} \tag{2.46}$$

where, $\boldsymbol{u}_i$ is a p-component vector with a "one" in position i and zeros elsewhere.

Now, the new restricted master program is formulated using the current master program basis and the new columns. Solve the new master program by the revised simplex method. Then, go back to step 1 and repeat the process.

Note that in the revised simplex method, the transformed entering column is calculated by $\boldsymbol{B}^{-1}\boldsymbol{p}$, where $\boldsymbol{B}^{-1}$ is the basis inverse. The simplex pivot operation is used to obtain a new basis inverse and a new vector of simplex multipliers. A matrix notation throughout the following example tableaus will point to the basis inverse.

2.5.3 Example of the Dantzig-Wolfe Decomposition.

Example 2.5

Minimize $z = -3x_1 - x_2 - y_1 - 4y_2$

St.
$$\begin{aligned}
2x_1 + x_2 + y_1 + 2y_2 &\le 25 \\
2x_1 + x_2 \qquad\qquad &\le 20 \\
x_1 + x_2 \qquad\qquad &\le 10 \\
2y_1 + y_2 &\le 5 \\
x_1, x_2, y_1, y_2 &\ge 0
\end{aligned}$$

This problem has one coupling constraint and two independent blocks. The feasible regions for these two blocks are described by:

$$\boldsymbol{S}_1 = \{\boldsymbol{x} \mid \boldsymbol{B}_1\boldsymbol{x} = \boldsymbol{b}_1, \boldsymbol{x} \ge 0\} \qquad \boldsymbol{S}_2 = \{\boldsymbol{y} \mid \boldsymbol{B}_2\boldsymbol{y} = \boldsymbol{b}_2, \boldsymbol{y} \ge 0\}$$

This problem can be decomposed in a number of ways, each leading to a different number of convexity constraints in the master program. Two convexity constraints will be included to make fullest use of the restricted master program.

Let $\boldsymbol{x}$ and $\boldsymbol{y}$ be the feasible solutions for block 1 and block 2, respectively, and $\boldsymbol{x} = \sum_i \alpha_i \boldsymbol{x}^i$, $\boldsymbol{y} = \sum_i \beta_i \boldsymbol{y}^i$ where $\boldsymbol{x}^i$ and $\boldsymbol{y}^i$ are extreme points of these blocks. The extreme points for blocks 1 and 2 are $\{(0,0),(10,0),(0,10)\}$ and $\{(0,0),(2.5,0),(0,5)\}$, respectively.

The master program then becomes:

$$\textit{Minimize} \qquad z = \sum_i (\boldsymbol{c}_1 \boldsymbol{x}^i)\alpha_i + \sum_i (\boldsymbol{c}_2 \boldsymbol{y}^j)\beta_i$$

$$\textit{St.} \qquad \begin{aligned} \sum_i (\boldsymbol{a}_1 \boldsymbol{x}^i)\alpha_i + \sum_i (\boldsymbol{a}_2 \boldsymbol{y}^i)\beta_i + s &= 35 \\ \sum_i \alpha_i &= 1 \\ \sum_i \beta_i &= 1 \\ \alpha_i \geq 0, \quad \beta_i \geq 0 & \end{aligned}$$

Where $\boldsymbol{c}_1 = (-3,-1), \boldsymbol{c}_2 = (-1,-4), \boldsymbol{a}_1 = (2,1), \boldsymbol{a}_2 = (1,2)$, and $s \geq 0$ is a slack variable. An initial basic feasible solution may be obtained by choosing $\boldsymbol{x}^1 = \boldsymbol{y}^1 = (0,0)$, $\alpha_1 = \beta_1 = 1$, $s = 35$.

Thus the initial revised simplex tableau is as follows:

Basic Variables	s	α_1	β_1	$-z$	constant
s	1	0	0	0	25
α_1	0	1	0	0	1
β_1	0	0	1	0	1
$-z$	0	0	0	1	0

Iteration 1. The negatives of the simplex multipliers for the master program are the first three elements in the bottom row of the above tableau and are currently all zero, $(\pi, \pi_{01}, \pi_{02}) = (0,0,0)$.

Thus the first set of subproblems are:

$$\begin{aligned} &\text{Minimize} \quad z_1 = c_1 x = -3x_1 - x_2 \\ &\text{St.} \\ &\quad 2x_1 + x_2 \le 20 \\ &\quad x_1 + x_2 \le 10 \\ &\quad x_1, x_2 \ge 0 \end{aligned}$$

$$\begin{aligned} &\text{Minimize} \quad z_2 = c_2 y = -y_1 - 4y_2 \\ &\text{St.} \\ &\quad 2y_1 + y_2 \le 5 \\ &\quad y_1, y_2 \ge 0 \end{aligned}$$

The optimal solution is:

$$x^2 = (10,0), \quad z_1^0 = -30, \quad y^2 = (0,5), \quad z_2^0 = -20$$

The minimal relative cost factors are:

$$\min \bar{f}_x = z_1^0 - \pi_{01} = -30 < 0$$

$$\min \bar{f}_y = z_2^0 - \pi_{02} = -20 < 0$$

Associated with each subproblem solution is a column of the master program. The columns corresponding to the current solution are:

$$\begin{pmatrix} a_1 x^2 \\ 1 \\ 0 \end{pmatrix} = \begin{pmatrix} 20 \\ 1 \\ 0 \end{pmatrix}, \quad \begin{pmatrix} a_2 y^2 \\ 0 \\ 1 \end{pmatrix} = \begin{pmatrix} 10 \\ 0 \\ 1 \end{pmatrix}$$

Where $c_1 x^2 = -30$ and $c_2 y^2 = -20$ are the associated cost coefficients.

The restricted master program is now formulated using the current master program basis plus the new columns:

$$\begin{aligned} &\text{Minimize} && -30\alpha_2 - 20\beta_2 \\ &\text{St.} && s + 20\alpha_2 + 10\beta_2 = 25 \\ &&& \alpha_1 + \alpha_2 = 1 \\ &&& \beta_1 + \beta_2 = 1 \\ &&& \alpha_1, \alpha_2, \beta_1, \beta_2 \ge 0 \end{aligned}$$

The revised Simplex method is used:

Relative cost factors					
Cycle	s	α_1	β_1	α_2	β_2
0	0	0	0	-30	-20
1	0	30	0	0	-20
2	2	-10	0	0	0
3	1.5	0	5	0	0

Basic Variables	s	α_1	β_1	$-z$	constant	Transformed entering column
	Cycle 0: α_2 *enters*, α_1 *leaves*					
s	1	0	0	0	25	20
α_1	0	1	0	0	1	(1)
β_1	0	0	1	0	1	0
$-z$	0	0	0	1	0	-30

Basic Variables	s	α_1	β_1	$-z$	constant	Transformed entering column
	Cycle 1: β_2 *enters*, s *leaves*					
s	1	−20	0	0	5	(10)
α_2	0	1	0	0	1	0
β_1	0	0	1	0	1	1
$-z$	0	30	0	1	30	-20

Basic Variables	s	α_1	β_1	$-z$	constant	Transformed entering column
	Cycle 2: α_1 *enters*, β_1 *leaves*					
β_2	.1	−2	0	0	.5	-2
α_2	0	1	0	0	1	1
β_1	−.1	2	1	0	.5	(2)
$-z$	2	-10	0	1	40	-10

Basic Variables	s	α_1	β_1	$-z$	constant
	Cycle 3: Optimal				
β_2	0	0	1	0	1
α_2	.05	0	−.5	0	.75
α_1	−.05	1	.5	0	.25
$-z$	1.5	0	5	1	42.5

The new solution to the primal problem is:

$$\mathbf{x} = \alpha_1 \mathbf{x}^1 + \alpha_2 \mathbf{x}^2 = 0.25(0,\ 0) + 0.75(10, 0) = (7.5, 0)$$

$$\mathbf{y} = \beta_2 \mathbf{y}^2 = 1(0,5) = (0,5)$$

with objective value -42.5 .

Iteration 2. The new simplex multipliers are found in the bottom row of the final tableau: $(\pi, \pi_{01}, \pi_{02}) = (-1.5, 0, -5)$

These are used to form new subproblem objectives:

$$z_1 = (\boldsymbol{c}_1 - \pi\boldsymbol{a}_1)\boldsymbol{x} = [(-3,-1) + 1.5(2,1)]\boldsymbol{x} = 0.5x_2$$
$$z_2 = (\boldsymbol{c}_2 - \pi\boldsymbol{a}_2)\boldsymbol{y} = [(-1,-4) + 1.5(1,2)]\boldsymbol{y} = 0.5y_1 - y_2$$

Minimizing these over the subproblem constraints yield:

$$\boldsymbol{x}^3 = (0,0) \ \ or \ \ (10,0), \qquad z_1^0 = 0$$
$$\boldsymbol{y}^3 = (0,5), \qquad z_2^0 = -5$$

with minimal relative cost factors:

$$min \ \bar{f}_x = z_1^0 - \pi_{01} = 0$$
$$min \ \bar{f}_y = z_2^0 - \pi_{02} = -5 + 5 = 0$$

Since these are non-negative, the current primal solution is optimal.

2.6 APPLICATION OF DANTZIG-WOLFE DECOMPOSITION TO POWER LOSS MINIMIZATION

2.6.1 Introduction. This topic will be used in Chapter 7 of this book. Redistribution of reactive power generation in a power system is necessary to improve the system voltage profiles and to minimize the real power losses. Reactive power distributions in the power system can be controlled by adjusting system control variables like transformer tap position, generator voltages as well as switchable VAR sources. These control variables have permissible upper and lower limits; any changes to these variables will have the effect of changing the voltage profile, reactive power output of generators and system losses. Hence, the loss minimization problem can be stated as: find the set of adjustments to control variables for minimizing power transmission losses, while satisfying the other constraints.

This is a nonlinear optimization problem. The objective function is the real power loss in the system and constraints are system operation limits. The goal is accomplished by re-allocating the power generation in the system like adjusting unit generators, condensers, shunt capacitor banks and tap transformers.

There are two possibilities for the solution of this nonlinear optimization problem. The first is to use a nonlinear optimization technique. The traditional nonlinear optimization techniques suffer from many drawbacks such as insecure convergence properties, large CPU time and complexity. The second alternative is to apply the linear programming approach which is a fast, reliable, versatile and sufficiently accurate method to solve the optimal reactive power dispatch problem for minimizing losses. Because of the superiority of the LP technique over the nonlinear techniques, we use LP in this section.

2.6.2 Loss Minimization Problem.

2.6.2.1 Objective function. The total real power loss P_L in Figure 2.2 is represented by the following function:

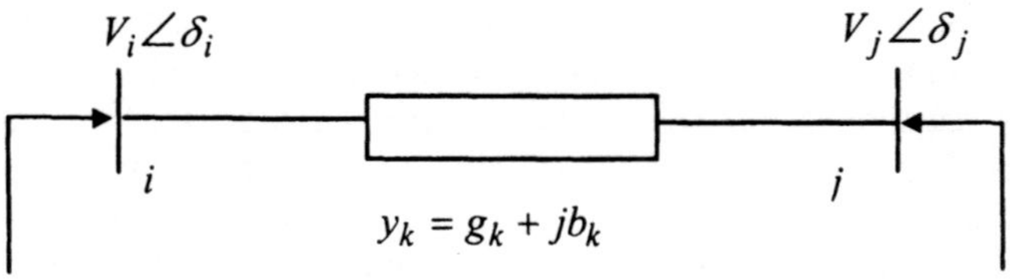

Figure 2.2 Transmission Line *k* Connecting Bus *i* and Bus *j*

$$P_L = \sum_{k=1}^{nls} g_k [V_i^2 + V_j^2 - 2V_iV_j\cos(\delta_i - \delta_j)] \tag{2.47}$$

In the incremental form,

$$\Delta P_L = \begin{pmatrix} \frac{\partial P_L}{\partial V_1} & \frac{\partial P_L}{\partial V_2} & \cdots\cdots & \frac{\partial P_L}{\partial V_n} \end{pmatrix} \begin{pmatrix} \Delta V_1 \\ \Delta V_2 \\ \cdot \\ \cdot \\ \Delta V_n \end{pmatrix} \tag{2.48}$$

or

$$\Delta P_L = \boldsymbol{M} \cdot \boldsymbol{\Delta V} \tag{2.49}$$

where

$$\frac{\partial P_L}{\partial V_i} = \sum_{k \in i} g_k [2V_i - 2V_j\cos(\delta_i - \delta_j)] \tag{2.50}$$

$$\frac{\partial P_L}{\partial V_j} = \sum_{k \in j} g_k [2V_j - 2V_i\cos(\delta_i - \delta_j)] \tag{2.51}$$

We will show in this section that the loss minimization formulation can be given as follows:

$$Min \ \ \Delta P_L = \boldsymbol{M}\,\Delta V$$

$$St. \ \ \Delta \boldsymbol{Q}^{min} \le \boldsymbol{J}_1''\Delta V \le \Delta \boldsymbol{Q}^{max}$$

$$\boldsymbol{J}_2''\,\Delta\,\boldsymbol{V} = \boldsymbol{0}$$

$$\Delta V^{min} \le \Delta V \le \Delta V^{max}$$

In the above equation, the first set of constraints represent busses connected to reactive power sources and the second set represents fixed load and junction buses. Bus voltage limits are given by the third set of constraints. This formulation encompasses reactive power injections by tap changing transforms and loads at various buses. The intriguing part of the above formulation is that the only variable is the bus voltage, and it does not require any matrix inversion for calculating bus voltages and transmission losses (as it is the case with other methods in the literature). In the following, we discuss the tap changing transformer and load models as a function of bus voltage.

2.6.2.2 Transformer model. The equivalent model of a tap changing transformer is shown in Figure 2.3.

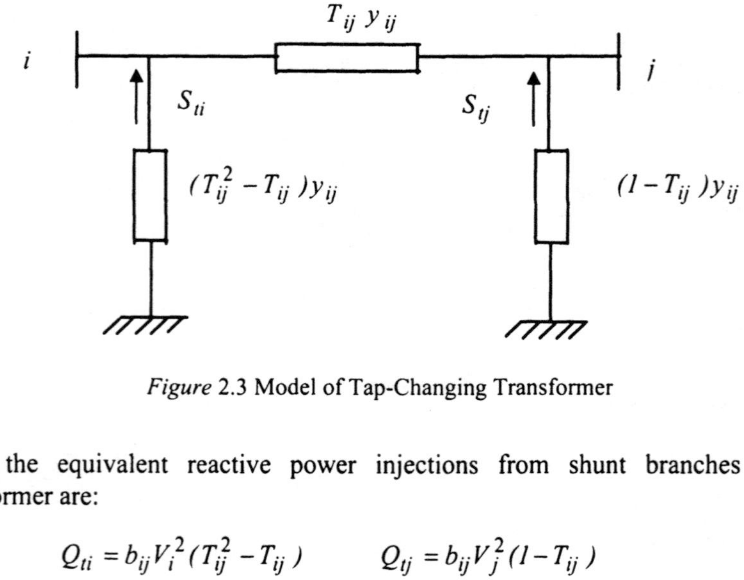

Figure 2.3 Model of Tap-Changing Transformer

Here, the equivalent reactive power injections from shunt branches of the transformer are:

$$Q_{ti} = b_{ij}V_i^2(T_{ij}^2 - T_{ij}) \qquad Q_{tj} = b_{ij}V_j^2(1 - T_{ij}) \tag{2.52}$$

The linearized reactive power injections are:

$$\Delta Q_{ti} = \Delta Q_{ti}^V + \Delta Q_{ti}^T = 2b_{ij}V_i(T_{ij}^2 - T_{ij})\Delta V_i + b_{ij}V_i^2(2T_{ij} - 1)\Delta T_{ij} \tag{2.53}$$

$$\Delta Q_{tj} = \Delta Q_{tj}^V + \Delta Q_{tj}^T = 2b_{ij}V_j(T_{ij}^2 - T_{ij})\Delta V_j - b_{ij}V_j^2\Delta T_{ij} \tag{2.54}$$

In forming the Y-bus matrix, shunt branches of the equivalent transformer model are not included in the diagonal elements because equivalent reactive power injections are used to simulate the effects of these branches on the system.

2.6.2.3 Load model. The effect of voltage variation on the power consumed by system loads depends on a great extent on the type of load that is being supplied by the power system. Reactive loads can be represented as a function of bus voltage magnitude as follows:

$$Q_{di} = Q_{dsi}\left(\frac{V_i}{V_{si}}\right)^{q_i} \tag{2.55}$$

Where, base case reactive power and voltage are denoted by Q_{dsi} and V_{si}, respectively. The characteristic of the load can be specified by the value of q_i which corresponds to 0 for constant power load, 1 for constant current load and 2 for constant impedance load. So, the variation of reactive power at the load with respect to changes of the bus voltage magnitude can be represented by:

$$\Delta Q_{di} = \frac{\partial Q_{di}}{\partial V_i}\Delta V_i = \frac{q_i Q_{di}}{V_{si}}\left(\frac{V_i}{V_{si}}\right)^{q_i - 1} \cdot \Delta V_i \tag{2.56}$$

Now, we incorporate these two models in the loss minimization equation.

2.6.2.4 Network model. For the tap changing transformer model, we consider the case where bus *i* is connected to generator and bus j is not, we have the following relationship between reactive power injections and bus voltage:

$$Q_i = Q_{gi} + Q_{ti} - Q_{di} = \sum_{j=1}^{N} V_i V_j Y_{ij} sin(\delta_i - \delta_j - \theta_{ij}) \tag{2.57}$$

$$Q_j = Q_{tj} - Q_{dj} = \sum_{k=1}^{N} V_j V_k Y_{jk} sin(\delta_j - \delta_k - \theta_{jk}) \tag{2.58}$$

The linearized form of equation (2.57) is:

$$\Delta Q_i = \Delta Q_{gi} + \Delta Q_{ti} - \Delta Q_{di} = [\boldsymbol{J}_i]\,\Delta \boldsymbol{V} \tag{2.59}$$

$$\Delta Q_j = \Delta Q_{tj} - \Delta Q_{dj} = [\boldsymbol{J}_j]\,\Delta \boldsymbol{V} \tag{2.60}$$

Where $\Delta \boldsymbol{V}$ is a vector of bus voltage change and:

$$J_{ij} = \frac{\partial Q_i}{\partial V_j} = V_i Y_{ij} sin(\delta_i - \delta_j - \theta_{ij}) \tag{2.61}$$

$$J_{ii} = \frac{\partial Q_i}{\partial V_i} = -V_i Y_{ii} \sin\theta_{ii} + \sum_{j=1}^{N} V_j Y_{ij} \sin(\delta_i - \delta_j - \theta_{ij}) \tag{2.62}$$

Substituting (2.53) and (2.56) in (2.58), we obtain the following equations:

$$\Delta Q_{gi} + \Delta Q_{ti}^T + \Delta Q_{ti}^V = [J_i'] \, \Delta V \tag{2.63}$$

$$\Delta Q_{tj}^T + \Delta Q_{tj}^V = [J_j'] \, \Delta V \tag{2.64}$$

where,

$$J_{ii}' = J_{ii} + \frac{q_i Q_{di}}{V_{si}} \left(\frac{V_i}{V_{si}}\right)^{q_i - 1} \tag{2.65}$$

$$J_{jj}' = J_{jj} + \frac{q_j Q_{dj}}{V_{sj}} \left(\frac{V_j}{V_{sj}}\right)^{q_j - 1} \tag{2.66}$$

$$J_{ij}' = J_{ij} \tag{2.67}$$

Substituting (2.53) and (2.54) in (2.64):

$$2b_{ij} V_j (1 - T_{ij}) \Delta V_j - b_{ij} V_j^2 \Delta T_{ij} = [J_j'] \, \Delta V \tag{2.68}$$

$$\Delta Q_{tj}^T = -b_{ij} V_j^2 \Delta T_{ij} = [J_j''] \, \Delta V \tag{2.69}$$

Where,

$$J_{jj}'' = J_{jj}' - 2b_{ij} V_j (1 - T_{ij}) \tag{2.70}$$

$$J_{jk}'' = J_{jk}' \qquad (j \neq k, \; k = 1, \ldots\ldots N) \tag{2.71}$$

From (2.69), we obtain:

$$\Delta T_{ij} = -\frac{1}{b_{ij} V_j^2} [J_j''] \, \Delta V \tag{2.72}$$

Putting (2.69) in (2.63), we get:

$$\Delta Q_{gi} = [J_i''] \, \Delta V \tag{2.73}$$

Where

$$J_{ii}'' = J_{ii}' - 2b_{ij} V_i (T_{ij}^2 - T_{ij}) + \frac{V_i^2 (2T_{ij} - 1)}{V_j^2} J_{ji}'' \tag{2.74}$$

$$J''_{ik} = J'_{ik} + \frac{V_i^2(2T_{ij} - 1)}{V_j^2} J''_{jk} \qquad (k \neq i, k = 1, \ldots\ldots. N) \tag{2.75}$$

2.6.2.5 Determination of limits. To formulate power loss constraints, the following two types of buses in the system have to be considered for reactive power injection limits:

(1) Buses with reactive power generation and/or connection to tap changing transformer terminals.

(2) Load and junction buses neither connected to reactive power generation nor to tap transformer terminals.

The limits of the constraint equations are formed according to the type of bus. We have three different types to consider in our case:

(a) Buses with reactive power sources which are connected to transformer terminals (bus i). The limits are:

$$\Delta Q_i^{max} = \Delta Q_{gi}^{max} = Q_{gi}^{max} - Q_{gi} \tag{2.76}$$

$$\Delta Q_i^{min} = \Delta Q_{gi}^{min} = Q_{gi}^{min} - Q_{gi} \tag{2.77}$$

(b) Buses which are connected to transformer terminals (bus j). The limits are:

$$\Delta Q_i^{max} = \Delta Q_{ti}^{max} \tag{2.78}$$

$$\Delta Q_i^{min} = \Delta Q_{ti}^{min} \tag{2.79}$$

(c) Junction and load buses, not connected to transformer and generator terminals. They represent the equality constrains.

Finally, the reactive optimal power flow formula for loss minimization is given as follows:

$$\begin{aligned} &Min \;\; \Delta P_L = \boldsymbol{M}\,\Delta \boldsymbol{V} \\ &St. \;\; \Delta \boldsymbol{Q}^{min} \le \boldsymbol{J}''_1 \Delta \boldsymbol{V} \le \Delta \boldsymbol{Q}^{max} \\ &\qquad \boldsymbol{J}''_2\, \Delta\, \boldsymbol{V} = \boldsymbol{0} \\ &\qquad \Delta \boldsymbol{V}^{min} \le \Delta \boldsymbol{V} \le \Delta \boldsymbol{V}^{max} \end{aligned} \tag{2.80}$$

In Section 2.6.3, we will discuss the application of the Dantzig-Wolfe decomposition method to the solution of (2-80). The subject discussed in this section will be utilized in Chapter 7 as we will discuss the short-term transmission line maintenance scheduling problem.

2.6.2.6 Digital solution algorithm. The steps in the digital solution are shown in Figure 2.4, and are summarized below:

Step 1. Calculate the initial load-flow solution.

Step 2. Formulate the Jacobian matrix J, and add load and tap changing effect to form J''.

Step 3. Linearize system losses. Formulate and solve the LP problem for the given objective function and the set of constraints.

Step 4. Monitor the values of (ΔQ_{step}, ΔV_{step}, ΔT_{step}) at each iteration. If any of the step sizes is beyond the limits, the value of the limit is used.

Step 4. Check whether the real power losses in the system are significantly different from that of previous iteration. If so, repeat the process, otherwise calculate all variable and stop.

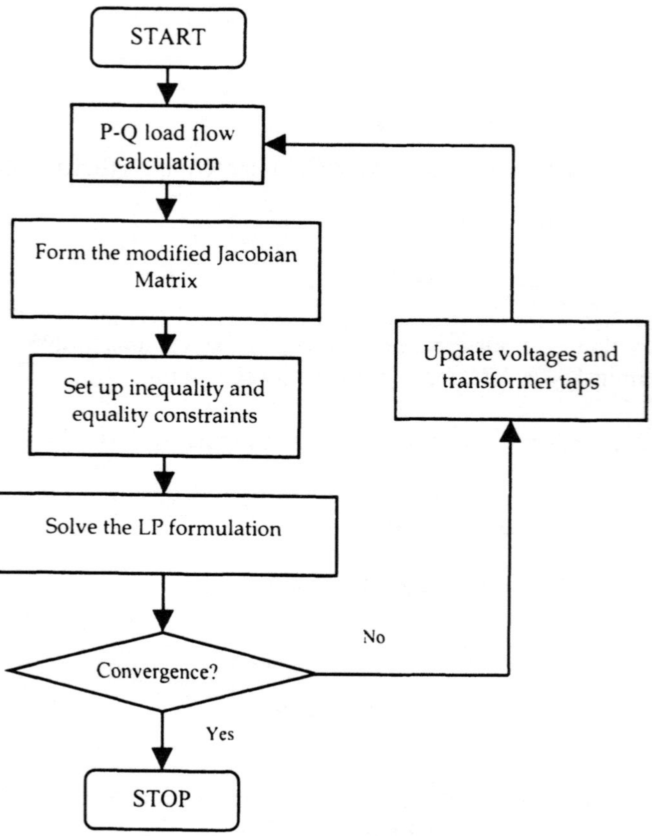

Figure 2.4 Flowchart for the Reactive OPF Solution Algorithm

2.6.3 Application of Dantzig-Wolfe. Since transmission lines are owned by separate entities, it is logical to consider the Dantzig-Wolfe decomposition method to solve (2-80). A Dantzig-Wolfe decomposition for this application is illustrated in Figure 2.5.

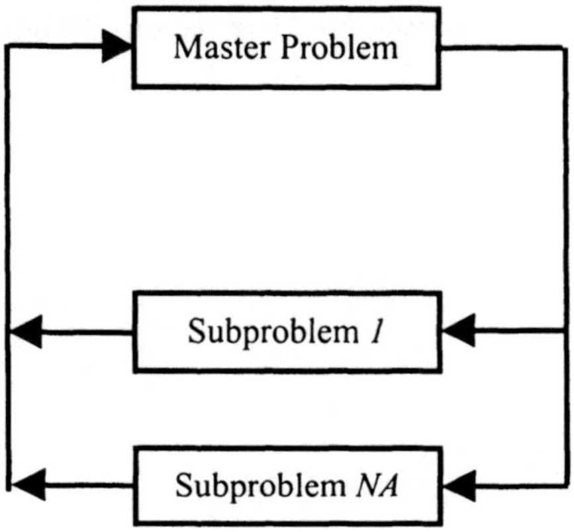

Figure 2.5 Iterative Procedure of Dantzig-Wolfe Method

Here, subproblems signify the role of TRANSCOs in a restructured system and the master problem represents the ISO. Each TRANSCO will try to minimize its losses and adjust its voltage profile while the ISO will coordinate the entire process. To apply the Dantzig-Wolfe decomposition, the reactive power optimization (2.80) is formulated in a block-angular form with *NA* TRANSCOs as follows:

$$
\begin{aligned}
& Min \sum_{n=1}^{NA} \boldsymbol{M}_n \cdot \Delta \boldsymbol{V}_n \\
& St. \sum_{n=1}^{NA} \boldsymbol{A}_n \cdot \Delta \boldsymbol{V}_n = \boldsymbol{b}_0 \\
& \boldsymbol{D}_n \cdot \Delta \boldsymbol{V}_n = \boldsymbol{b}_n \qquad (2.81) \\
& \Delta \boldsymbol{V}_n^{min} \le \Delta \boldsymbol{V}_n \le \Delta \boldsymbol{V}_n^{max} \\
& n = 1, \ldots, NA
\end{aligned}
$$

where, $\boldsymbol{A}_n$ is a coefficient matrix which corresponds to linking constraints of area *n*, and $\boldsymbol{D}_n$ is a coefficient matrix which corresponds to the subproblem *n* (in area *n*).

The system equations are decomposed according to the TRANSCOs' properties in power systems. Areas are identified by buses which are controlled by individual

TRANSCOs. The constraints that link buses within a specific area will form a subproblem. Buses within an area with links to other areas will form linking constraints [DEE91].

2.6.3.1 Implementation of the Dantzig-Wolfe decomposition method.

Here we use the modified IEEE 30-bus system as an example with three TRANSCOs in Figure 2.6. Line and bus data for this 30-bus system are given in Tables 2.3 and 2.4.

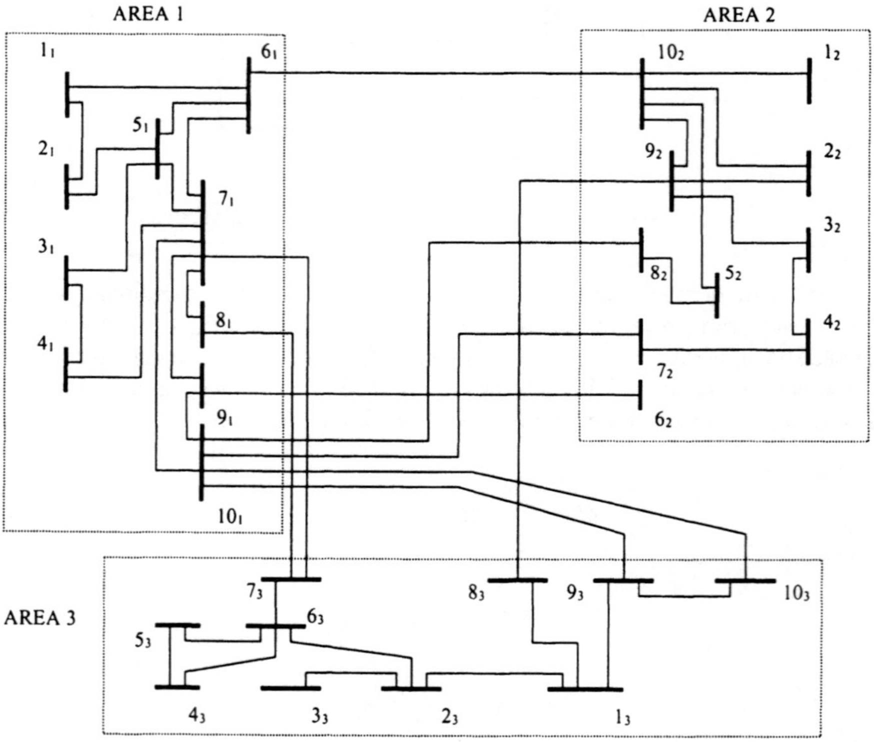

Figure 2.6 Schematic Diagram of the Decomposed 30-Bus System

In general, let us consider a *nb*-bus power system with a total of *NA* TRANSCOs. In TRANSCO *n*, the total number of buses is N_n. Then, according to (2.81), the block-angular form of the modified IEEE 30-bus system is written as follows:

$$Minimize \sum_{n=1}^{3} \sum_{i=1}^{N_n} \left(\frac{\partial P_L}{\partial V_i}\right)_n (\Delta V_i)_n \tag{2.82}$$

Subject to,

Coupling constraints consisting of the following:
I.C. representing buses: $6_1, 8_1, 9_1, 10_1, 6_2, 8_2, 10_2, 8_3$
E.C. representing buses: $7_1, 7_2, 9_2, 7_3, 9_3, 10_3$

AREA 1
I.C. 1,2,3,4,5

AREA 2
I.C. 1,3,4
E.C. 2.5

AREA 3
I.C. 1,4,6
E.C. 2.3.5

where I.C. stands for inequality constraints, and E.C. denotes equality constraints.

Here, each TRANSCO is controlling all buses within its area (intra-area buses) that are not linked to other areas (TRANSCOs), and the ISO is controlling the remaining (inter-area) buses that are linked to other areas (TRANSCOs). For example, in area 1, TRANSCO 1 is controlling buses 1-5 and the ISO is controlling buses 6-10. In area 2, TRANSCO 2 is controlling buses 1-5 and the ISO is controlling buses 6-10. Likewise, in area 3, TRANSCO 3 is controlling buses 1-6 and the ISO is controlling buses 7-10.

Table 2.2 shows the loss minimization results. According to the proposed method, losses have dropped from 10.24 to 6.51 MW for the entire 30-bus system. The computation took 14 iterations and 1.5 seconds on an IBM personal computer. It is noted in our study that the loss minimization results are equally impressive for a large-scale power system.

Table 2.2 Loss Minimization Results Applied to the IEEE 30-Bus System

Iteration number	Iterations (MW)
Initial state	10.24
	10.01
	9.66
	9.38
	9.08
	8.69
	8.24
	8.01
	7.72
	7.43
	7.09
	6.63
	6.85
	6.64
	6.51
Total no. of iterations	14
CPU time per iteration	1.5 sec

Table 2.3 Line Data for the IEEE 30-Bus System

Line Number	From - To	R (p.u.)	X (p.u.)	Tap Ratio (T)*
1	1 - 2	0.0192	0.0575	
2	1 - 3	0.0452	0.1852	0.9610
3	2 - 4	0.0570	0.1737	0.9560
4	3 - 4	0.0132	0.0379	
5	2 - 5	0.0472	0.1983	
6	2 - 6	0.0581	0.1763	
7	4 - 6	0.0119	0.0414	
8	5 - 7	0.0460	0.1160	
9	6 - 7	0.0267	0.0820	0.9650
10	6 - 8	0.0120	0.0420	
11	6 - 9	0.0000	0.2080	
12	6 - 10	0.0000	0.5560	
13	9 - 11	0.0000	0.2080	
14	9 - 10	0.0000	0.1100	
15	4 - 12	0.0000	0.2560	0.9530
16	12 - 13	0.0000	0.1400	0.9700
17	12 - 14	0.1231	0.2559	
18	12 - 15	0.0662	0.1304	0.9850
19	12 - 16	0.0945	0.1987	
20	14 - 15	0.2210	0.1997	
21	16 - 17	0.0824	0.1932	
22	15 - 18	0.1070	0.2185	0.9635
23	18 - 19	0.0639	0.1292	
24	19 - 20	0.0340	0.0680	
25	10 - 20	0.0936	0.2090	
26	10 - 17	0.0324	0.0845	
27	10 - 21	0.0348	0.0749	
28	10 - 22	0.0727	0.1499	
29	21 - 22	0.0116	0.0236	
30	15 - 23	0.1000	0.2020	0.9590
31	22 - 24	0.1150	0.1790	
32	23 - 24	0.1320	0.2700	
33	24 - 25	0.1885	0.3292	
34	25 - 26	0.2544	0.3800	0.9655
35	25 - 27	0.1093	0.2087	
36	28 - 27	0.0000	0.3960	
37	27 - 29	0.2198	0.4153	
38	27 - 30	0.3202	0.6027	
39	29 - 30	0.2399	0.4533	0.9810
40	8 - 28	0.0636	0.2000	
41	6 - 28	0.0169	0.0599	

* For all transformers in the system, $0.9 \leq T \leq 1.1$

Table 2.4 Bus Data for the IEEE 30-Bus System

Bus Number*	V**	δ	Q^{min}	Q^{max}
G1	1.060	0.00	-0.200	0.000
G2	1.030	-2.20	-0.200	0.200
3	0.976	-4.00		
4	0.967	-4.80		
G5	0.988	-7.50	-0.150	0.150
6	0.960	-5.60		
7	0.948	-7.00		
G8	0.959	-5.80	-0.150	0.150
9	0.955	-7.00		
10	0.914	-9.30		
G11	0.975	-4.50	-0.100	0.100
12	0.949	-7.80		
G13	0.936	-5.00	-0.150	0.150
14	0.935	-9.00		
15	0.933	-9.30		
16	0.921	-8.80		
C17	0.902	-9.50	-0.050	0.050
C18	0.930	-10.30	0.000	0.055
19	0.889	-10.60		
20	0.894	-10.30		
21	0.900	-10.00		
22	0.900	-10.00		
C23	0.929	-10.10	-0.050	0.055
24	0.888	-10.60		
25	0.901	-11.00		
26	0.881	-11.70		
C27	0.919	-10.80	-0.055	0.055
28	0.945	-6.20		
29	0.884	-12.30		
30	0.875	-13.50		

* G: Generator Bus   C: Capacitor Bus

** $0.00 \le V \le 1.06$ and for all other buses $0.90 \le V \le 1.10$

CHAPTER III

LONG-TERM GENERATION MAINTENANCE SCHEDULING

3.1 INTRODUCTION

Long-term generation maintenance scheduling determines time intervals, within a one to two year planning horizon, when generating units in a GENCO can be taken off-line for planned preventive maintenance. Long-term generation maintenance schedule should minimize the total operating cost in a GENCO while the GENCO interacts with the ISO to meet the energy demand, reliability and other system requirements. Since we are dealing with independent entities in a restructured system, we apply Benders decomposition to include the ISO's constraints in the long-term generation maintenance scheduling problem. Figure 3.1 depicts the solution methodology.

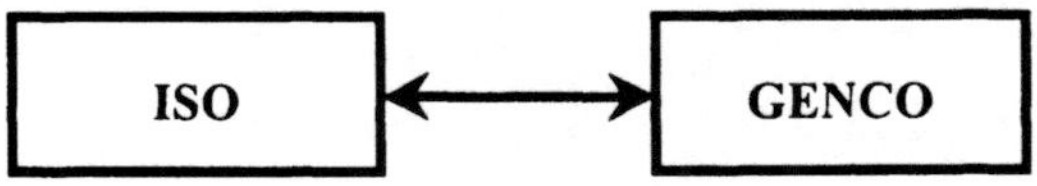

Figure 3.1 Interactions between the ISO and GENCO

In general, there are two possible approaches that could be employed by GENCOs for long-term generation maintenance scheduling. The first one is based on fictitious costs to penalize deviations from a preferred generation maintenance schedule. The second approach is based on identifying maintenance windows (time intervals) in which the preferred maintenance schedule is represented within windows and the objective will be to minimize the real maintenance cost instead of a fictitious cost. Throughout this book, we use the maintenance window approach.

The organization of the chapter is as follows. Section 3.2 describes the formulation of maintenance scheduling with ISO constraints. The solution methodology is discusses in Section 3.3. The network model used in the proposed method, the maintenance sub-problem and operating sub-problem are also discussed in Section 3.3. The results for a simple example and the application of the proposed method to IEEE-RTS [Rel79] are presented in Sections 3.4 and 3.5. In section 3.6, we discuss a include the reliability constraints and forced outages of generating units in calculating planned outages. Sections 3.7 and 3.8 present an example and then more elaborated case studies for the probabilistic model. In Section 3.9, we include additional constraints for fuel consumption and emission in scheduling generating unit maintenance. Sections 3.10 and 3.11 present the solution method for this problem. The case studies for this comprehensive model are given in Section 3.12.

3.2 MAINTENANCE PROBLEM FORMULATION

Mathematically, long-term generation maintenance scheduling can be formulated as follows:

$$Min \sum_{t}\sum_{i}\left\{C_{it}(1-x_{it})+c_{it}g_{it}\right\}$$

S.T. maintenance constraints:

$$\begin{array}{lll} x_{it}=1 & for\ t\le e_i\ or\ t\ge l_i+d_i & \\ x_{it}=0 & for\ s_i\le t\le s_i+d_i & (i) \\ x_{it}=0\ or\ 1 & for\ e_i\le t\le l_i & \end{array}$$

$$\begin{array}{lll} 1.\ crew\ availability & 3.\ seasonal\ limitations & \\ 2.\ resources\ availability & 4.\ desirable\ schedule & (ii) \end{array}$$

system constraints:

$$\begin{array}{llll} \boldsymbol{Sf}+\boldsymbol{g}+\boldsymbol{r}=\boldsymbol{d} & \forall t & (iii) & \\ \boldsymbol{g}\le\bar{\boldsymbol{g}}.\boldsymbol{x} & \forall t & (iv) & (3.1) \\ \boldsymbol{r}\le\boldsymbol{d} & \forall t & (v) & \\ |\boldsymbol{f}|\le\bar{\boldsymbol{f}} & \forall t & (vi) & \\ \sum_i r_{it}\le\varepsilon & \forall t & (vii) & \end{array}$$

Eq. (3.1) corresponds to a mixed-integer programming problem since x_{it} is an integer variable and g_{it} is continuous. The objective of (3.1) is to minimize the total maintenance and production costs in a GENCO over the operational planning period. The first term of objective function (3.1) is the maintenance cost of generators and the second is the energy production cost. Constraints (i) represent the maintenance window stated in terms of maintenance variables (s_i). The generation unit maintenance may not be scheduled before its earliest period (e_i) or after its latest period allowed for maintenance (e.g., l_i+d_i). The set of constraints (ii) represents crew and resources availability, seasonal limitations, desirable schedules, as well as other constraints such as fuel and emission constraints.

Seasonal limitations may be incorporated in e_i and l_i values of constraint (i). If we consider, for example, that units 1, 2 and 3 are to be maintained simultaneously, the set of constraints would be:

$$x_{1t}+x_{2t}+x_{3t}=3 \qquad \text{or} \qquad x_{1t}+x_{2t}+x_{3t}=0$$

If we consider that in each maintenance area within a GENCO we have limited resources and crew available, the set of constraints will be stated as follows:

$$\sum_{i \in A} \sigma_{mi}(1 - x_{it}) \leq Z_{mt} \tag{3.2}$$

In the case of representing a resource constraint, Z_{mt} would be the amount of resource m available in area A for each time t and σ_{mi} would be a percentage of this resource required for unit i. In the case of a crew constraint, Z_t would be the number of maintenance crew in area ω and σ_i would be a percentage of this number required for maintenance of unit i.

The set of system constraints (iii)-(vii), which represent the peak load balance, transmission flow limits and allowable unserved energy will be checked by the ISO.

In order to calculate the maintenance schedule, it is essential that numerous and complex constraints which limit the choice of scheduling times are incorporated into the solution method. The constraints in the maintenance scheduling problem are categorized as coupling and decoupling.

3.2.1 Coupling Constraints. The first of these coupling constraints is the requirement that generating units be overhauled regularly. This procedure is incorporated periodicity by specifying min/max times that a generating unit may run without maintenance. The required time for overhaul is generally known and, hence, the number of weeks that a machine is "down" is pre-determined. It is assumed here that there is little flexibility in manpower usage that varies the required time for maintenance. Furthermore, only a limited number of generating units may be serviced at once due to a limited manpower. The available crews can be split into geographical and organizational types. The number of crews in each type required at each stage of overhaul of each unit is specified.

3.2.2 Decoupling Constraints. Network constraints in each time period are considered as decoupling constraints; the network can be modeled as either the transportation model or a linearized power flow model. We use transportation model to represent system operation limits, peak load balance equation, and generation and line capacity limits.

3.3 SOLUTION METHODOLOGY

The generating unit maintenance problem is decomposed into a master problem (GENCO) and a set of independent operation sub-problems (ISO). The master problem, which in this model is an integer-programming problem, is solved to generate a trial solution for maintenance schedule decision variables. This master problem is a relaxation of the original problem as it contains only a subset of constraints. Its optimal objective is a lower bound on the optimal objective of the original problem. Once x_{it} variables are fixed, the ISO sub-problem will solve a set of independent sub-problems, one for each time period t, since there are no constraints across time periods. At each iteration, the solution of sub-problems generates dual multipliers which measure the change in either the production cost or the reliability resulting from marginal changes in the trial maintenance scheduling. These dual multipliers are used to form one or more constraints (i.e., Benders cuts) which are added to the master problem for the next iteration. The process continues until a feasible solution is found whose cost is sufficiently close to the lower bound. This process is shown in Figure 3.2.

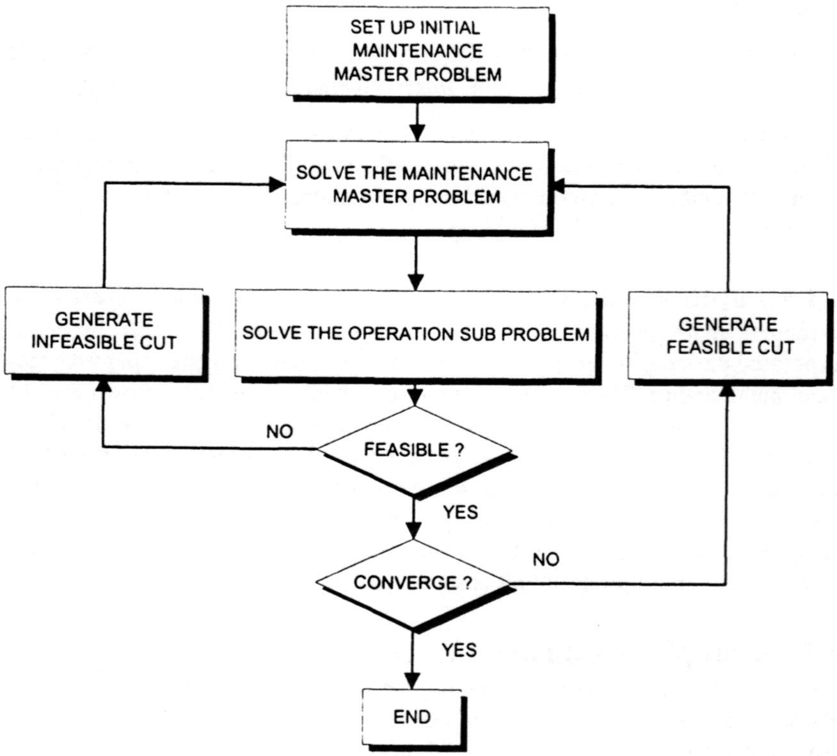

Figure 3.2 Maintenance Scheduling Decomposition

If we let X denote the vector of maintenance variables $\{x_{it}\}$, Ω represent the set of maintenance schedules for which constraints (iii)-(vii) are satisfied in all periods t, then we define the operation cost w_t as:

$$w_t = \sum_i c_{it} g_{it}$$

Here (3.1) can be written as:

$$Min\sum_t\sum_i C_{it}(1-x_{it})+\sum_t Min\{w_t|(iii)\text{-}\{vii)\}$$

$S.T.$

maintenance constraints :

$$x_{it}=1 \qquad for\ t\le e_i\ or\ \ t\ge l_i+d_i$$

$$x_{it}=0 \qquad for\ s_i\le t\le s_i+d_i \qquad (i) \qquad (3.3)$$

$$x_{it}=0\ or\ 1 \qquad for\ e_i\le t\le l_i$$

1. crew availability   *3. seasonal limitations*

2. resources availability   *4. desirable schedule*   *(ii)*

$$X\in\Omega$$

If the tth sub-problem was a linear program, it could be replaced by its dual as in the standard Benders decomposition. The Lagrangian dual of the tth sub-problem is given by

$$L_t=max_{\kappa,\pi,\gamma,\zeta,\mu\ge 0}\{L_t(\kappa,\pi,\gamma,\zeta,\mu)\} \tag{3.4}$$

where $L_t(\kappa,\pi,\gamma,\zeta,\mu)$ is the Lagrangian function and κ,π,γ,ζ and μ are multipliers of constraints (iii)-(vii).

$$L_t(\kappa,\pi,\gamma,\zeta,\mu)=min_{g\ge 0}\left\{\begin{array}{l} w_t+\sum_i\kappa_{it}\left(\sum_k\left(S_{ik}f_{kt}\right)+g_{it}+r_{it}-d_{it}\right)+ \\ \sum_i\pi_{it}\left(g_{it}-\bar{g}_i.x_{it}\right)+\sum_i\gamma_{it}\left(r_{it}-d_{it}\right)+ \\ \sum_k\varsigma_{kt}\left(\left|f_{kt}\right|-\bar{f}_k\right)+\mu_t\left(\left(\sum_i r_{it}\right)-\varepsilon\right)\end{array}\right\} \tag{3.5}$$

The tth sub-problem is then replaced by L_t, and (3.3) is rewritten as:

$$Min \sum_t \sum_i C_{it}(1-x_{it}) + L_t$$

$S.T.$

maintenance constraints :

$$
\begin{array}{lll}
x_{it} = 1 & for\ t \le e_i\ or\ t \ge l_i + d_i & \\
x_{it} = 0 & for\ s_i \le t \le s_i + d_i & (i) \qquad (3.6) \\
x_{it} = 0\ or\ 1 & for\ e_i \le t \le l_i &
\end{array}
$$

$$
\begin{array}{lll}
1.crew availabilty & 3.seasonal limitations & \\
2.resources availabilty & 4.preschedule & (ii)
\end{array}
$$

$$X \in \Omega$$

To ensure $X \in \Omega$, the maintenance schedule must provide a sufficient capacity reserve while minimizing the cost of operation. The tth sub-problem is feasible if and only if the optimal value of the following problem is less than ε

$$
\begin{aligned}
& Min \sum_i r_{it} \\
& S.T.\ \boldsymbol{Sf} + \boldsymbol{g} + \boldsymbol{r} = \boldsymbol{d} \\
& \qquad \boldsymbol{g} \le \bar{\boldsymbol{g}}.\boldsymbol{x} \qquad (3.7) \\
& \qquad \boldsymbol{r} \le \boldsymbol{d} \\
& \qquad |\boldsymbol{f}| \le \bar{\boldsymbol{f}}
\end{aligned}
$$

Its Lagrangian dual is

$$max_{\nu,\lambda,\tau,\eta \ge 0} U_t(\nu,\lambda,\tau,\eta)$$

where U_t (ν,λ,τ,η) is the following Lagrangian function and ν,λ,τ and η are multipliers of constraints (iii)-(vii).

$$
U_t(\nu,\lambda,\tau,\eta) = min_{g \ge 0} \left\{
\begin{array}{l}
\sum_i r_{it} + \sum_i \nu_{it}\left(\left(\sum_k S_{ik} f_{kt}\right) + g_{it} + r_{it} - d_{it}\right) + \\
\sum_i \lambda_{it}(g_{it} - \bar{g}_i.x_{it}) + \sum_i \tau_{it}(r_{it} - d_{it}) \\
+ \sum_k \eta_{kt}(|f_{kt}| - \bar{f}_k)
\end{array}
\right\}
$$

We then arrive at the generalized Benders master problem:

$$Min \quad z$$

$$S.T.$$

$$z \geq \sum_t \sum_i C_{it}(1 - x_{it}) + \sum_t L_t(\kappa, \pi, \gamma, \zeta, \mu) \quad \text{for all} \quad \kappa, \pi, \gamma, \zeta, \mu \geq 0$$

$$\sum_t U_t(\nu, \lambda, \tau, \eta) \leq \varepsilon \quad \text{for all} \quad \nu, \lambda, \tau, \eta \geq 0$$

maintenance constraints :

$$\begin{array}{lll} x_{it} = 1 & \text{for } t \leq e_i \text{ or } t \geq l_i + d_i & \\ x_{it} = 0 & \text{for } s_i \leq t \leq s_i + d_i & (i) \\ x_{it} = 0 \text{ or } 1 & \text{for } e_i \leq t \leq l_i & \end{array} \tag{3.8}$$

1. crew availability *3. seasonal limitations*

2. resources availability *4. desirable schedule* *(ii)*

The initial maintenance master problem is formulated as follow:

$$Min \quad z$$

$$S.T. \quad z \geq \sum_t \sum_i \{C_{it}(1 - x_{it})\}$$

maintenance constraints :

$$\begin{array}{ll} x_{it} = 1 & \text{for } t \leq e_i \text{ or } t \geq l_i + d_i \\ x_{it} = 0 & \text{for } s_i \leq t \leq s_i + d_i \\ x_{it} = 0 \text{ or } 1 & \text{for } e_i \leq t \leq l_i \end{array} \tag{3.9}$$

1. crew availability *3. seasonal limitations*

2. resources availability *4. desirable schedule*

3.3.1 ISO Sub-Problem. The sub-problem will not have any solutions if the unserved energy cannot be kept above a desired level. If a sub-problem is infeasible, then an infeasibility cut will be generated. For each infeasible sub-problem resulting from the nth trial solution of the master problem, the infeasibility cut is of the form

$$\sum_i r_{it}^n + \sum_i \lambda_{it}^n \, \overline{g}_i \, (x_{it}^n - x_{it}) \leq \varepsilon \tag{3.10}$$

Multiplier λ_{it}^{n} may be interpreted as the marginal decrease in unserved energy with a 1 MW increase in generation, given the nth trial maintenance schedule. The infeasibility cuts (3.10) would eliminate those maintenance values which could not be scheduled.

If the sub-problem is feasible, then the fuel cost for period t, w_t, would depend on the utilization of available generating units to satisfy load constraints in each time period subject to maintaining the unserved energy above a certain level. Thus the generation cost in period t is expressed as:

$$\begin{aligned} w_t &= Min \sum_i c_{it} g_{it} \\ S.T.\ & \boldsymbol{Sf} + \boldsymbol{g} + \boldsymbol{r} = \boldsymbol{d} \\ & \boldsymbol{g} \le \bar{\boldsymbol{g}} \boldsymbol{x}^n \\ & \boldsymbol{r} \le \boldsymbol{d} \\ & |\boldsymbol{f}| \le \bar{\boldsymbol{f}} \\ & \sum_i r_{it} \le \varepsilon \end{aligned} \tag{3.11}$$

The solution of this sub-problem is not complicated since we know generators and transmissions that are available during period t. The feasibility cut is of the form:

$$z \ge \sum_t \left(w_t^n + \sum_i \left(C_{it}(1 - x_{it}^n) + \pi_{it}^n \bar{g}_i (x_{it}^n - x_{it}) \right) \right) \tag{3.12}$$

where w_t^n is the expected fuel cost for period t associated with the nth trial solution. The multiplier π_{it}^n may be interpreted as the marginal cost associated with 1 MW decrease in the generation capacity, given the nth trial maintenance schedule. The cost cut (3.12) will tend to increase the lower bounds obtained in each successive maintenance sub-problem solution.

3.3.2 Revised Maintenance Master Problem.

The revised maintenance master problem would minimize the maintenance cost subject to maintenance constraints as well as feasibility and infeasibility cuts from the operation sub-problems. If all sub-problems are feasible then their solutions would yield a set of dual multipliers from which a feasibility cut is constructed. If one or more operation sub-problems are infeasible then for each infeasible sub-problem an infeasibility cut would be generated. Hence,

$$Min \quad z$$

$$S.T.\, z \geq \sum_t \sum_i \left\{ C_{it}(1 - x_{it}) \right\}$$

- *maintenance constraint :*

$$x_{it} = 1 \qquad for \;\; t \leq e_i \;\; or \;\; t \geq l_i + d_i$$

$$x_{it} = 0 \qquad for \;\; s_i \leq t \leq s_i + d_i$$

$$x_{it} = 0 \; or \; 1 \qquad for \;\; e_i \leq t \leq l_i$$

1. crew availability   *3. seasonal limitations*

2. resources availability   *4. desirable schedule*

- *feasibility and infeasibility cuts from previous iterations*
- *if all sub - problems are feasible then the feasible cut is :*

$$z \geq \sum_t \left(w_t^n + \sum_i C_{it}(1 - x_{it}^n) + \pi_{it}^n \, \overline{g}_i (x_{it}^n - x_{it}) \right)$$

- *if one or more sub - problem is infeasible then the infeasible cuts are :*

$$\sum_i r_{it}^n + \sum_i \lambda_{it}^n \, \overline{g}_i \, (x_{it}^n - x_{it}) \leq \varepsilon \qquad \forall t \in infeasible \; sub\text{-}problem$$

$$x \in 0,1 \tag{3.13}$$

where n is the current number of iterations

λ^n, π^n are the multiplier vectors at n^{th} iteration

The important feature of the Benders decomposition is the availability of upper and lower bounds to the optimal solution in each iteration. These bounds are used as an effective convergence criterion. The convergence criterion is

$$\frac{2(w - z)}{(w + z)} \leq \Delta \tag{3.14}$$

Here, there is a trade-off between the value of Δ and the number of iterations. The smaller the Δ, the larger the number of iterations will be. For some cases, the problem may not converge if Δ is too small. From our experience, setting Δ at 1% would be good enough.

3.4 EXAMPLE

We use a three-bus system as an example. Later we consider the generating units in IEEE-RTS [Appendix A] for maintenance scheduling. The maximum unserved energy requirement (ε) is 0 p.u., and generator and line data, in p.u., are given in Tables 3.1 and 3.2 with load data depicted in Figure 3.3. We perform maintenance on at least one generator. We assume the study period only has one time interval and loads are constant during the study period.

First, we solve the initial maintenance master problem.

<u>Maintenance master problem iteration 1:</u>

Min z

$$S.T.\ 300*(1-x_1)+200*(1-x_2)+100*(1-x_3) \le z$$
$$x_1+x_2+x_3 \le 2$$
$$x_1\le 1,\ x_2\le 1,\ x_3\le 1$$

The solution is: $x_1=1, x_2=1, x_3=0, z=100$.

Table 3.1 Generator Data for 3-bus System

Unit	Min Cap. (p.u.)	Max Cap. (p.u.)	Cost ($)	Maint. Cost/Unit ($)
1	0.5	2.5	10 g_1	300
2	0.6	2.5	10 g_2	200
3	0.6	3.0	10 g_3	100

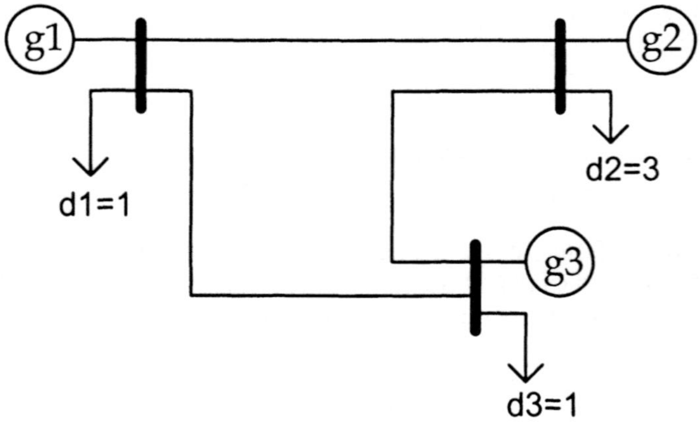

Figure 3.3 Three-Bus System Example

Table 3.2 Line Data for 3-bus System

Line	Ω/line	# of lines	Capacity/line (p.u.)
1-2	0.2	2	0.25
2-3	0.25	2	0.5
1-3	0.4	2	0.25

Operation sub-problem iteration 1:

We check the feasibility of the operation sub-problem given the first trial of the maintenance schedule. The feasibility check is as follows:

$Min\ r_1+r_2+r_3$

$S.T.\ -f_{12}-f_{13}+g_1+r_1=1$	Load balance at bus 1
$-f_{23}+f_{12}+g_2+r_2=3$	Load balance at bus 2
$f_{13}+f_{23}+g_3+r_3=1$	Load balance at bus 3
$0.5 \le g_1 \le 2.5$	Generator 1 limit
$0.6 \le g_2 \le 2.5$	Generator 2 limit
$0.0 \le g_3 \le 0.0$	Generator 3 limit
$-2*0.25 \le f_{12} \le 2*0.25$	Line 1-2 flow limit
$-2*0.25 \le f_{13} \le 2*0.25$	Line 1-3 flow limit
$-2*0.5 \le f_{23} \le 2*0.5$	Line 2-3 flow limit

The primal solution of the feasibility check is given as : r=0.5 g_1=2 g_2=2.5 g_3=0 f_{12}=0.5 f_{13}=0.5 f_{23}=0.

The dual price of the operation sub-problem is: $\lambda_{g1}=0$ $\lambda_{g2}=1$ $\lambda_{g3}=1$.
The above LP solution is infeasible, since $r_1+r_2+r_3 \ge 0$. So, the generation cost is set arbitrarily to 1000 because the solution is infeasible. The infeasibility cut is as follows: $0.5 + 1*2.5*(1-x_2) + 1*3*(0-x_3) \le 0$.

Maintenance master problem iteration 2:

$Min\ z$

$S.T.\ 300*(1-x_1)+200*(1-x_2)+100*(1-x_3) \le z$

$0.5 + 1*2.5*(1-x_2) + 1*3*(0-x_3) \le 0.5$

$x_1+x_2+x_3 \le 2$

$x_1 \le 1,\ x_2 \le 1,\ x_3 \le 1$

The solution is: x_1= 1 x_2=0 x_3=1 z=200.

Operation sub-problem iteration 2:

The feasibility check is as follows:

$Min\ r_1+r_2+r_3$

$S.T.\ -f_{12}-f_{13}+g_1+r_1=1$	Load balance at bus 1
$-f_{23}+f_{12}+g_2+r_2=3$	Load balance at bus 2
$f_{13}+f_{23}+g_3+r_3=1$	Load balance at bus 3
$0.5 \le g_1 \le 2.5$	Generator 1 limit
$0.0 \le g_2 \le 0.0$	Generator 2 limit
$0.6 \le g_3 \le 3.0$	Generator 3 limit
$-2*0.25 \le f_{12} \le 2*0.25$	Line 1-2 flow limit
$-2*0.25 \le f_{13} \le 2*0.25$	Line 1-3 flow limit
$-2*0.5 \le f_{23} \le 2*0.5$	Line 2-3 flow limit

The primal solution of the feasibility check is given as : r=1.5 g_1=1.5 g_2=0 g_3=2 f_{12}=0.5 f_{13}=0 f_{23}= -1.

The dual price of the operation sub-problem is: $\lambda_{g1}=0 \quad \lambda_{g2}=1 \quad \lambda_{g3}=0$.

The above LP solution is infeasible, since $r_1+r_2+r_3\geq 0$. The generation cost is set arbitrarily to 1000 because the solution is infeasible. The infeasibility cut is as follows: $1.5 + 1*2.5*(1-x_2) \leq 0$.

Maintenance master problem iteration 3:

$$
\begin{aligned}
&Min \quad z \\
&S.T. \quad 300*(1-x_1)+200*(1-xg_2)+100*(1-xg_3) \leq z \\
&\qquad 0.5 + 1*2.5*(1-x_2) + 1*3*(0-x_3) \leq 0.5 \\
&\qquad 1.5 + 1*2.5*(1-x_2) \leq 0 \\
&\qquad x_1+x_2+x_3 \leq 2 \\
&\qquad x_1\leq 1, \quad x_2\leq 1, \quad x_3\leq 1
\end{aligned}
$$

The solution is: $x_1=0 \quad x_2=1 \quad x_3=1 \quad z=300$.

Given the trial maintenance schedule in iteration 3, we apply the feasibility check as before which gives r=0. This means the trail schedule is feasible now.

Operation sub-problem iteration 3:

The feasible sub-problem is as follows:

$$
\begin{aligned}
&Min \quad w=10*g_2+10*g_3+300 \\
&S.T. \quad -f_{12}-f_{13}+g_1+r_1=1 && \text{Load balance at bus 1} \\
&\qquad -f_{23}+f_{12}+g_2+r_2=3 && \text{Load balance at bus 2} \\
&\qquad f_{13}+f_{23}+g_3+r_3=1 && \text{Load balance at bus 3} \\
&\qquad 0.0 \leq g_1 \leq 0.0 && \text{Generator 1 limit} \\
&\qquad 0.6 \leq g_2 \leq 2.5 && \text{Generator 2 limit} \\
&\qquad 0.6 \leq g_3 \leq 3.0 && \text{Generator 3 limit} \\
&\qquad -2*0.25 \leq f_{12} \leq 2*0.25 && \text{Line 1-2 flow limit} \\
&\qquad -2*0.25 \leq f_{13} \leq 2*0.25 && \text{Line 1-3 flow limit} \\
&\qquad -2*0.5 \leq f_{23} \leq 2*0.5 && \text{Line 2-3 flow limit} \\
&\qquad r_1+r_2+r_3 \leq 0
\end{aligned}
$$

The primal solution is: $w=350 \quad g_2=2.5 \quad g_3=2.5 \quad f_{12}=-0.5 \quad f_{13}=-0.5 \quad f_{23}=-1$.
The dual price of the operation sub-problem is: $\pi_{g1}=0 \quad \pi_{g2}=0 \quad \pi_{g3}=0$.
All dual prices of the sub-problem are zero since the generators have the same costs of \$10/p.u.

3.5 CASE STUDIES

We apply the proposed method to IEEE-RTS [Appendix A]. This system is made of 32 generating units, 20 demand sides, 23 buses and 38 transmission lines. The generating units encompass a single GENCO. A three-month study period of summer weeks 18-29 is considered. It is presumed that a few generating facilities in a particular area of the GENCO will need maintenance within the study period. The coverage of the designated maintenance area is from buses 1 through 10. Table 3.3 gives the generating units locations within the maintenance area.

Table 3.3 Generating Units Considered For Maintenance

Unit	Capacity (MW)	Bus
1	2x76	1
2	2x76	2
3	1x100	7
4	2x100	7
5	2x20	1

Table 3.4 gives the operating characteristics of generating units in this area. The average fuel price is \$3.00/MBtu for Oil #2, \$1.2/MBtu for coal and \$2.3/MBtu for Oil #6. Detailed system data for transmission lines, generators and loads can be seen in Appendix A. The unit maintenance cost C_{it}, is the maintenance cost of Table 3.4 multiply by penalty factors in Table 3.5.

Table 3.4 Generating Unit Operating Cost Data

Size MW	Fuel	Heat Rate Btu/kWh	Maintenance Cost		Maintenance	
			Fixed 10^3 \$/Yr.	Variable \$/kW/Yr.	Window (weeks)	Duration (weeks)
20	Oil #2	14500	6.0	5.0	18-29	2
76	Coal	12000	760.0	0.9	18-29	2
100	Oil #6	10000	850.0	0.8	18-29	2

Table 3.5 Unit Maintenance Cost Penalty Factors

Week	Factor	Week	Factor	Week	Factors
18	1.000	22	1.333	26	1.667
19	1.083	23	1.417	27	1.750
20	1.167	24	1.500	28	1.833
21	1.250	25	1.583	29	1.917

The minimization of production and maintenance costs is used as the objective function. The unserved energy in each week is limited to 1% of the total energy. The results of the following test cases show the effect of the ISO (network constraints) on maintenance schedule and generating units loading points.

- Case 0: GENCO constraints are considered only.
- Case 1: ISO constraints are considered as well.

In Case 0, the problem is the classical unit maintenance schedule in a GENCO. Case 0 is converged in two iterations. In the first iteration, sub-problems are infeasible in all periods. In the second iteration, sub-problems are feasible and the final cost is given. The maximum transmission flow over the three-month study period is shown in Table 3.6 in which each line flow, except for those of lines 1-3 and 11-14, is within limits at most time periods.

Table 3.6 Overflow in Transmission Lines (12-week horizon, $\in$=1% of load)

Case 0		
Line Between buses	Rating MVA	Peak Flow MVA
1-3	208	215
11-14	600	610

Case 1 studies the effect of transmission flow limits on maintenance schedule. The imposed transmission flow limits increase the cost of production. Table 3.7 shows the change in maintenance cost and production cost over the study period, indicating a shift from generating units that use inexpensive fuel to those with more expensive fuel and inefficient units.

The corresponding generator loading points of the weekly load are shown in Table 3.8. In this table, the 12-week horizon is between weeks 18-29; weeks up and down in Table 3.8 refer to periods in which units are used for supplying the load.

Table 3.7 Total Cost for Generating Unit (12-week horizon, $\in$=1% of load)

Case	Cost ($\$10^6$)
0	36.62694
1	36.67142

Table 3.8 Effect of Network Constraints on Generating Units Maintenance (12-week horizon, $\in$=1% of load)

Case	Unit	Weeks on Maintenance	Weeks Up	Weeks Down
0	1	24,25	18-23,26-29	-
	2	27,28	18-26,29	-
	3	20,21	18,19,22-29	-
	4	18,19	20,21,24,25,27,28	22,23,26,29
	5	22,23	-	18-21,24-29
1	1	23,24	18-22,25-29	-
	2	28,29	18-27	-
	3	20,21	18,19,22-29	-
	4	18,19	21-24,28,29	20,25-27
	5	27,28	23,24	18-22,25,26,29

In practice, generating units are loaded in decreasing order of operating costs. With transmission flow limitations, available generating units in one time period may become less attractive as compared to those in other time periods. This in turn increases the generating unit operating cost and affects the loading of units as shown in Table 3.8. Here, the inefficient generating unit 5 has to be brought on-line in weeks 23-24 to supply the generation deficit.

The maintenance of IEEE-RTS was tested for a 52-week period as shown in Table 3.9. Since unit maintenance is not confined to the 12-week period, the total cost is reduced as given in Table 3.10 and a better distribution of risk is achieved. Here, the unit maintenance schedule is shifted to lower peak loads: the maintenance of unit 1 is shifted from weeks 23-24 to weeks 33-34. Also those of units 2, 3, 4 and 5 are shifted from weeks 28-29 to weeks 38-39, weeks 20-21 to 10-11, weeks 18-19 to 12-13, weeks 27-28 to 36-37, respectively.

Table 3.9 Generating Unit Maintenance in Case 1 (52-week horizon, $\in$=1% of load)

Unit	Weeks on Maintenance
1	33,34
2	38,39
3	10,11
4	12,13
5	36,37

Table 3.10 Total Cost for Generating Unit (52-week horizon, $\in$=1% of load)

Case	Cost ($\$10^6$)
12-week	36.67142
52-week	36.65084

Here, we considered one GENCO in the case studies. However, in the case of multiple GENCOs, the cuts will be divided linearly among entities. Next we consider the above formulation in a probabilistic state in which the forced outages of components are included in maintenance scheduling.

3.6 PROBABILISTIC FORMULATION

The need to represent operating conditions uncertainties in system planning is widely recognized. The generation maintenance scheduling, which evaluates the availability of generating units in the planning period, takes into account equipment outages and load variations (composite reliability evaluation). We use a transportation model for the ISO to represent the system operation limits, peak load balance equation, and generation and line flow limits.

The generation maintenance with equipment outage possibilities is formulated as follows:

$$Min \sum_t \sum_i \left\{ C_{it}(1 - x_{it}) \right\} + E\left\{ \sum_t \sum_i c_{it} g_{it} \right\}$$

S.T.

maintenance constraints :

$$x_{it} = 1 \qquad for\ t \le e_i\ or\ t \ge l_i + d_i$$

$$x_{it} = 0 \qquad for\ s_i \le t \le s_i + d_i \qquad (i)$$

$$x_{it} \in \{0,1\} \qquad for\ e_i \le t \le l_i$$

1. crew availability *3. seasonal limitations*

2. resources availability *4. desirable schedule* *(ii)*

system constraints :

$$\begin{aligned}
&\boldsymbol{Sf}+\boldsymbol{g}+\boldsymbol{r}=\boldsymbol{d}(\varphi) && \forall t && (iii) \\
&\boldsymbol{g}\le \bar{\boldsymbol{g}}(\varphi).\boldsymbol{x} && \forall t && (iv) \\
&\boldsymbol{r}\le \boldsymbol{d}(\varphi) && \forall t && (v) \\
&|\boldsymbol{f}|\le \bar{\boldsymbol{f}}(\varphi) && \forall t && (vi) \\
&E\left\{\sum_i r_{it}\right\}\le \varepsilon && \forall t && (vii)
\end{aligned} \tag{3.15}$$

The initial maintenance master problem is formulated as follow:

$$\begin{aligned}
&Min \quad z \\
&S.T. \quad z \ge \sum_t \sum_i \left\{C_{it}(1-x_{it})\right\}
\end{aligned}$$

maintenance constraints :

$$\begin{aligned}
&x_{it}=1 && for\ t\le e_i\ or\ t\ge l_i+d_i \\
&x_{it}=0 && for\ s_i\le t\le s_i+d_i \\
&x_{it}\in\{0,1\} && for\ e_i\le t\le l_i
\end{aligned} \tag{3.16}$$

1. crew availability *3. seasonal limitations*

2. resources availability *4. desirable schedule*

The feasibility sub-problem is as follows:

$$\begin{aligned}
&Min\ E\left\{\sum_i r_{it}\right\} \\
&S.T.\ \boldsymbol{Sf}+\boldsymbol{g}+\boldsymbol{r}=\boldsymbol{d}(\varphi) \\
&\qquad \boldsymbol{g}\le \bar{\boldsymbol{g}}(\varphi).\boldsymbol{x}^n \qquad (dual\ \ variable\ is\ \lambda) \\
&\qquad \boldsymbol{r}\le \boldsymbol{d}(\varphi) \\
&\qquad |\boldsymbol{f}|\le \bar{\boldsymbol{f}}(\varphi)
\end{aligned} \tag{3.17}$$

For each infeasible sub-problem resulting from the nth solution of the master problem, the Benders cut is associated with a deterministic, instead of a probabilistic, operation sub-problem. The infeasibility cut is:

$$E\left\{\sum_{i} r_{it}^{n}\right\} + \sum_{i} \lambda_{it}^{n} \bar{g}_i (x_{it}^{n} - x_{it}) \leq \varepsilon \tag{3.18}$$

If all sub-problems are feasible, then the generation cost in period t will be expressed as:

$$\begin{aligned} & w_t = Min\ E\left\{\sum_{i} c_{it} g_{it}\right\} \\ & S.T.\ \boldsymbol{Sf} + \boldsymbol{g} + \boldsymbol{r} = \boldsymbol{d}(\varphi) \\ & \quad \boldsymbol{g} \leq \bar{\boldsymbol{g}}(\varphi).\boldsymbol{x}^n \qquad (dual\ variable\ is\ \pi) \\ & \quad \boldsymbol{r} \leq \boldsymbol{d}(\varphi) \\ & \quad |\boldsymbol{f}| \leq \bar{\boldsymbol{f}}(\varphi) \\ & \quad \sum_{i} r_{it} \leq \varepsilon(\varphi) \end{aligned} \tag{3.19}$$

The feasible cut is of the form:

$$z \geq \sum_{t} \left\{ w_t^n + \sum_{i} \left(C_{it} \bar{g}_i (1 - x_{it}^{n}) + \pi_{it}^{n} \bar{g}_i (x_{it}^{n} - x_{it}) \right) \right\} \tag{3.20}$$

The calculation of (3.17) and (3.19) involves the reliability evaluation of a composite system. The procedure is as follows:

1. Select a system state φ, i.e., define load levels, equipment availability, operating conditions, etc.
2. Calculate (3.17) for the selected state, i.e., verify whether the specific configuration of generators and transmission lines is able to supply the specific load without violating system limits.
3. Update the estimated of production cost or load shedding expectation.
4. Return to step 1 if the accuracy is unacceptable.

In order to allow a fast algorithm and efficient calculations, we use the simultaneous decomposition simulation for all generator and transmission states and intervals [Den92]. The revised master problem that consists of feasible and infeasible Benders cuts is as follows:

$$Min \quad z$$

$$S.T. \quad z \geq \sum_t \sum_i \left\{ C_{it}(1 - x_{it}) \right\}$$

- *maintenance constraints :*

$$x_{it} = 1 \qquad for \; t \leq e_i \; or \; t \geq l_i + d_i$$

$$x_{it} = 0 \qquad for \; s_i \leq t \leq s_i + d_i$$

$$x_{it} \in \{0,1\} \qquad for \; e_i \leq t \leq l_i$$

1. crew availability   *3. seasonal limitations*

2. resources availability   *4. desirable schedule*

- *feasibility and infeasibility cuts from previous iterations*
- *if all subproblems are feasible then the feasibilty cut is :*

$$z \geq \sum_t \left\{ w_t^n + \sum_i \left(C_{it} \bar{g}_i (1 - x_{it}^n) + \pi_{it}^n \, \bar{g}_i (x_{it}^n - x_{it}) \right) \right\}$$

- *if one or more subproblems are infeasible then the infeasiblity cut is :* (3.21)

$$E \left\{ \sum_i r_{it}^n \right\} + \sum_i \lambda_{it}^n \, \bar{g}_i \, (x_{it}^n - x_{it}) \leq 0$$

$\forall t \in$ *infeasible subproblem*

$xg \in 0,1$

where: n is the current number of iterations,

π^n, λ^n are multiplier vectors at the nth iteration.

3.7 EXAMPLE

We now include forced outage rates of generators in the three-bus system of Section 3.4. For convenience, in this sample study, all lines are assumed to be perfectly reliable. The forced outrage rate (FOR) of each generator is given in Table 3.11. The system reliability requirement (ε) enforced by the ISO is 0.5 p.u. The problem is to perform maintenance on one generator within the study period. We assume the study period has only one time interval and loads are constant during the study period.

Table 3.11 Generation Unit Data

Unit	Min Capacity (p.u.)	Max Capacity (p.u.)	Operation Cost ($)	Maint.Cost/Unit ($/Unit)	FOR
1	0.5	2.5	10 g_1	300	0.05
2	0.6	2.5	10 g_2	200	0.1
3	0.6	3.0	10 g_3	100	0.02

First, we solve the initial maintenance master problem:

Maintenance master problem iteration 1:

$$\begin{aligned}
&Min \quad z\\
&S.T. \quad 300*(1-x_1)+200*(1-x_2)+100*(1-x_3) \le z\\
&\qquad x_1+x_2+x_3 \le 2\\
&\qquad x_1 \le 1, \quad x_2 \le 1, \quad x_3 \le 1
\end{aligned}$$

The solution is: $x_1 = 1$, $x_2=1$, $x_3=0$, $z=100$.

Operation sub-problem iteration 1:

Then, we check the feasibility of operation sub-problem given the first trial of maintenance schedule. The first state space formulation of feasibility check is as follows:

$$\begin{aligned}
&Min \quad r_1+r_2+r_3\\
&S.T. \quad -f_{12}-f_{13}+g_1+r_1=1 && \text{Load balance at bus 1}\\
&\qquad -f_{23}+f_{12}+g_2+r_2=3 && \text{Load balance at bus 2}\\
&\qquad f_{13}+f_{23}+g_3+r_3=1 && \text{Load balance at bus 3}\\
&\qquad 0.5 \le g_1 \le 2.5 && \text{Generator 1 limit}\\
&\qquad 0.6 \le g_2 \le 2.5 && \text{Generator 2 limit}\\
&\qquad 0.0 \le g_3 \le 0.0 && \text{Generator 3 limit}\\
&\qquad -2*0.25 \le f_{12} \le 2*0.25 && \text{Line 1-2 flow limit}\\
&\qquad -2*0.25 \le f_{13} \le 2*0.25 && \text{Line 1-3 flow limit}\\
&\qquad -2*0.5 \le f_{23} \le 2*0.5 && \text{Line 2-3 flow limit}
\end{aligned}$$

The primal solutions for all state spaces in the feasibility check are given in Table 3.12.

Table 3.12 Feasibility Check State Spaces

φ		Prob.	***R***	Prob. X ***r***
g_1	g_2			
2.5	2.5	0.855	0.5	0.4275
0	2.5	0.045	2.5	0.1125
2.5	0	0.095	3	0.2850
0	0	0.005	5	0.0250
				$E\{r\}$=0.85

To find λ_{g1}, let increase the capacity of g_1 by 1 p.u. then the primal solutions for all state spaces in the feasibility check are given in Table 3.13.

Table 3.13 Feasibility Check State Spaces

φ		Prob.	r	Prob. X r
g_1	g_2			
3.5	2.5	0.855	0.5	0.4275
0	2.5	0.045	2.5	0.1125
3.5	0	0.095	3	0.2850
0	0	0.005	5	0.0250

$E\{r\}$=0.85

λ_{g1} = 0.85 – 0.85 = 0.00.

To find λ_{g2}, let increase the capacity of g_2 by 1 p.u. then the primal solutions for all state spaces in the feasibility check are given in Table 3.14.

Table 3.14 Feasibility Check State Spaces

φ		Prob.	r	Prob. x r
g_1	g_2			
2.5	3.5	0.855	0	0.0000
0	3.5	0.045	1.5	0.0675
2.5	0	0.095	3	0.285
0	0	0.005	5	0.025

$E\{r\}$=0.3775

λ_{g2} = 0.85 – 0.3775 = 0.4725.

To find λ_{g3}, let increase the capacity of g_3 by 1p.u. then the primal solutions for all state spaces in the feasibility check are given in Table 3.15.

Table 3.15 Feasibility Check State Spaces

φ			Prob.	R	Prob. x r
g_3	g_1	g_2			
1	2.5	2.5	0.8379	0	0.00000
1	0	2.5	0.0441	1.5	0.06615
1	2.5	0	0.0931	2	0.18620
1	0	0	0.0049	4	0.00196
0	2.5	2.5	0.0171	0.5	0.00855
0	0	2.5	0.0009	2.5	0.00225
0	2.5	0	0.0019	3	0.00570
0	0	0	0.0001	5	0.00050

$E\{r\}$=0.28895

λ_{g3} = 0.85 – 0.28895 = 0.56105.

The dual price of the operation sub-problem: λ_{g1}=0 λ_{g2} = 0.4725 λ_{g3}= 0.56105.

The above LP solution is infeasible, since $E\{r_1+r_2+r_3\}>0.5$. The cost is set arbitrarily to w=1000 because the solution is infeasible. The infeasible cut is as follows:

$0.85 + 0.4725*2.5*(1-x_2) + 0.56105*3*(0 - x_3) < 0.5$

The solution of the maintenance master problem for the second iteration is:

$x_1 = 1 \quad x_2 = 0 \quad x_3 = 1 \quad z = 500.$

Given the trial maintenance schedule in iteration 2, the feasibility check gives $E\{r\}=0.4225$ which is less than $\varepsilon=0.5$. The primal solution for all state spaces of the feasible sub-problem is given in Table 3.16.

Table 3.16 Feasible Sub-problem State Spaces

φ			Prob.	Cost	Prob. x cost
R	g_2	g_3			
0	2.5	2.5	0.882	50	47.025
3	0	2.5	0.098	25	2.450
2.5	2.5	0	0.018	25	0.450
5	0	0	0.002	0	0.000

$E\{10*g_2+10*g_3\} = 49.925$

w = 49.925 +300 = 349.925

To find π_{g2}, let increase the capacity of g_2 by 1 p.u. then the primal solutions for all state given in Table 3.17.

Table 3.17 Feasible Sub-problem State Spaces

φ			Prob.	Cost	Prob. x cost
r	g_2	g_3			
0	3.5	2.5	0.882	50	47.025
3	0	2.5	0.098	25	2.450
1.5	3.5	0	0.018	35	0.630
5	0	0	0.002	0	0.000

$E\{10*g_2+10*g_3\} = 50.105$

$\pi_{g1} = 49.925 - 50.105 = -0.18$

To find π_{g3}, let increase capacity of g_3 by 1 p.u. then the primal solutions for all state given in Table 3.18.

Table 3.18 Feasible Sub-problem State Spaces

φ			Prob.	Cost	Prob. x cost
r	g_2	g_3			
0	2.5	3.5	0.882	50	47.025
3	0	3.5	0.098	25	2.450
2.5	2.5	0	0.018	25	0.450
5	0	0	0.002	0	0.000

$E\{10*g_2+10*g_3\} = 49.925$

π_{g3} = 49.925 - 49.925 = 0.

The dual price of the operation sub-problem is: $\pi_{g1} = 0$ $\pi_{g2} = -0.98$ $\pi_{g3} = 0$

The feasible cut for the third iteration is:

$z \geq. 300*(1-x_1)+ 200*(1-x_2)+ 100*(1-x_3)+ 349.925-0.18*2.5*(1-x_2)$

The solution of the maintenance master problem for the third iteration is:

$x_1 = 0$ $x_2=1$ $x_3=1$ $z=349.925$.

We stop here since $z=w$, which means the cost, is equal to the lower bound.

3.8 CASE STUDIES

We repeat the proposed study for IEEE-RTS given in Section 3.5, except now we consider generation and transmission outages in Case2. Details of generation and transmission outage rates can be found in Appendix A. Expected Energy Not Supplied (EENS) in each week is limited to 1% of the total energy. The results of the following test case show the effect of network constraints (ISO) on maintenance schedule and unit loading points.

The imposed generation outages increase the cost of operation. Table 3.19 shows the change in operating cost over the study period, indicating a shift from units that use inexpensive fuel to those with more expensive fuel and inefficient units. Forced outages affect the maintenance schedule of unit 2, and cause the scheduling of inefficient units longer. The inefficient unit 5 is brought on-line during weeks 20 to 24 to supply the generation deficit.

Table 3.19 Total Cost for Generation Maintenance (12-week horizon, ε=1% of load)

Case	Cost (10^6 \$)
0	36.62694
1	36.67142
2	36.69147

The test system in Case 2 was applied to a 52-week horizon. Table 3.21 shows the maintenance schedule for generating units. Since units and lines are not forced to be on maintenance within the limited 12 weeks, the maintenance cost is reduced as seen in Table 3.22 and a better distribution of risk is achieved. In this case, generating units maintenance schedule is shifted to lower peak loads, i.e., maintenance of unit 1 is shifted from weeks 23-24 to weeks 34-35. Also maintenance of units 2, 3, 4 and 5 are shifted from weeks 28-29 to weeks 38-39, weeks 18-19 to 11-12, weeks 20-21 to 12-13, and weeks 27-28 to 38-39, respectively.

Table 3.20 Effect of Forced Outages on Generation Maintenance Scheduling (12-week horizon, ε=1% of load)

Case	Unit	Weeks on Maintenance	Weeks Up	Weeks Down
0	1	24,25	18-23,26-29	-
	2	27,28	18-26,29	-
	3	20,21	18,19,22-29	-
	4	18,19	20,21,24,25,27,28	22,23,26,29
	5	22,23	-	18-21,24-29
1	1	23,24	18-22,25-29	-
	2	28,29	18-27	-
	3	20,21	18,19,22-29	-
	4	18,19	21-24,28,29	20,25-27
	5	27,28	23,24	18-22,25,26,29
2	1	23,24	18-22,25-29	-
	2	28,29	18-27	-
	3	18,19	20-29	-
	4	20,21	18,19,23,24,28,29	22,25-29
	5	27,28	20-24	18-19,25,26,29

Table 3.21 Generating Unit Maintenance in Case 2 (52-week horizon, ε=1% of load)

Unit	Weeks on Maintenance
1	34,35
2	38,39
3	11,12
4	12,13
5	38,39

Table 3.22 Total Cost for Generation Maintenance

Case	Cost (10^6 \$)
12-week	36.69147
52-week	36.67934

3.9 FUEL AND EMISSION CONSTRAINTS

A comprehensive long–term generation maintenance scheduling must include fuel and inventory costs, operation and the maintenance costs, SO_2 allowances that are used, and the cost of purchasing energy from other GENCOs in its formulation.

In this section, we extend the study performed in the previous section to encompass fuel consumption and emission constraints in generation maintenance scheduling.

If each unit uses only one type of fuel and has a single supplier with an unrestricted fuel supply, the maintenance problem will be reduced to the problem in Section 3.6. When emission and additional fuel constraints are included, the problem becomes considerably more complex. Further complications would result when multiple fuel contracts would supply various stockpiles, which in turn supply one ore more of the generating units.

In an optimal maintenance schedule, the allocation of fuel to generating units must satisfy all contractual obligations and meet all limits on emission. The solution will specify off-line generating units for maintenance in each week, fuel allocation to on-line units, and SO_2 allowances consumed by on-line units.

Accounting for interactions between fuel restrictions, allowable emission and ISO constraints will greatly increase the complexity of the problem, resulting in a large optimization problem. The long-term generation maintenance scheduling problem has discrete decision variables related to maintenance scheduling and continuos variables representing fuel allocations and utilization levels of generating units. The corresponding objective for an operational planning period is given as,

$$Min \sum_t \left\{ \sum_i C_{it}(1-x_{it}) \right\} + \sum_i \sum_m \sum_t c_{mt}\phi_{imt} + \sum_t \chi_t E\left\{ \sum_i r_{it} \right\} \tag{3.22}$$

The first term in (3.22) is the maintenance cost of generators, the second is the energy production cost and the third is cost of energy purchased from outside. The production cost is a probabilistic optimization which takes into account the derated capacity of each generating unit. The sets of constraints are given as follows.

3.9.1 Maintenance and ISO Constraints. The constraints are the same as (3.15).

3.9.2 System Emission Limit. The system emission constraints are formulated as follows:

$$\begin{aligned} &\sum_t \sum_i H_i(E\{g_{it}\})x_{it} \le EMS \\ &\sum_t \sum_{i \in A} H_i(E\{g_{it}\})x_{it} \le EMA \end{aligned} \tag{3.23}$$

The two primary power plant emissions are *SO2* and *NOx*, given in our model as a function of unit generation. The emission can be modeled as either a linear or a quadratic function. We use a linear function in our case. However, in the case of a quadratic function, a piece-wise linear emission function will be adopted. (3.23) represents emission caps for certain areas of the system and the total system emission.

3.9.3 Fuel Constraints.

The fuel constraints are formulated as follows:

$$\sum_{m} \phi_{imt} \le K_i x_{it} \qquad \textit{for } \forall i \textit{ and } \forall t$$

$$\sum_{i} \phi_{imt} \le W_{mt} \qquad \textit{for } \forall m \textit{ and } \forall t$$

$$\sum_{i} \sum_{t \in month\ j} \phi_{imt} \le W_{mj} \qquad \textit{for } \forall m \textit{ and } \forall month\ j \tag{3.24}$$

$$\sum_{i} \sum_{t \in year\ v} \phi_{imt} \le W_{mv} \qquad \textit{for } \forall m \textit{ and } \forall year\ v$$

$$E\{g_{it}\} \le \sum_{m} \frac{\phi_{imt}}{h_{im}} \qquad \textit{for } \forall i \textit{ and } \forall t$$

The first constraint in (3.24) links the maintenance variable x_{it} to the fuel allocation variable ϕ_{imt}, while the remaining constraints relate to the fuel allocation variable. The first two constraints in (3.24) indicate that multiple fuels may supply a unit and a fuel contract may supply several units. Here, the long-term generation maintenance scheduling function observes restrictions which arise due to contractual agreements with fuel suppliers. These restrictions include the maximum amount of fuel that can be delivered during a month and a year, represented in the third and fourth constraints of (3.24). The last constraint is used to relate the generation level g_{it} to the fuel allocation ϕ_{imt}.

Maintenance and fuel constraints which are handled by the GENCO are considered as coupling constraints. ISO constraints are considered as decoupling constraints.

3.10 SOLUTION METHODOLOGY

In the case of a single fuel supplying a generating unit, the fuel price at a unit is known a priori and the single decomposition method in Section 3.6 can be used to handle fuel constraints at a unit. In a complex fuel allocation problem, where multiple fuels would supply a unit or a fuel contract supplies several units, the calculation of fuel price at a unit is part of the optimization process. In such a case, a double decomposition may be necessary to accommodate the general fuel allocation problem. Figure 3.4 depicts the process.

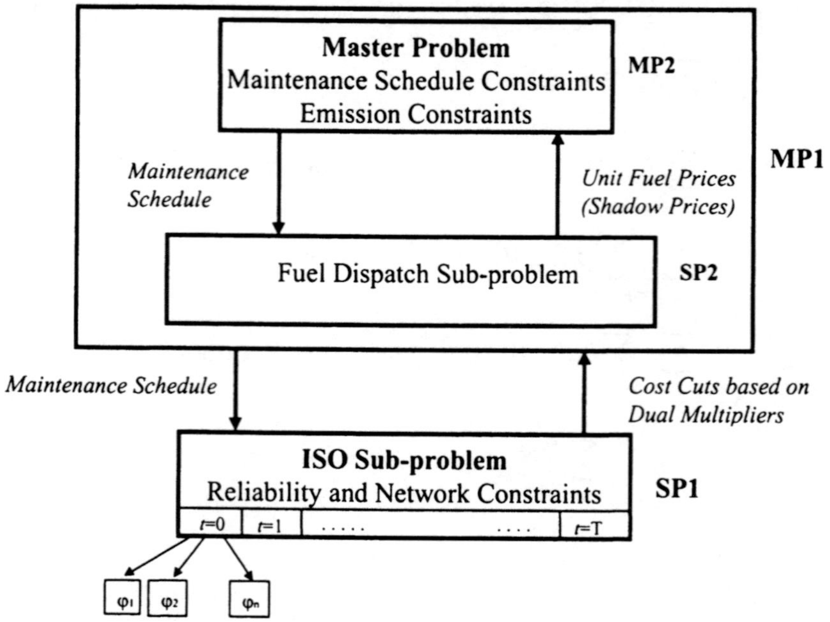

Figure 3.4 Long-term Decomposition with Fuel and Emission

In this figure, the problem is decomposed into several sections. MP1 represents the GENCO and SP1 is the operation subproblem representing the ISO. Once x_{it} variables are fixed by MP1, the resulting ISO sub-problem SP1 is treated as a set of independent sub-problems, one for each time period t, since there are no constraints across time periods. The set of ISO sub-problems in SP1 is then solved using the fixed maintenance schedule obtained from the solution of MP1. At each iteration, sub-problem SP1 calculates the unit generation level and the cost of purchasing energy while satisfying network constraints. SP1 also generates dual multipliers, which measure the change in unserved energy or energy cost resulting from marginal changes in the maintenance schedule. These dual multipliers are used to form one or more constraints (Benders cuts) which are added to the master problem MP1 for the next iteration. The process continues until a feasible solution is found whose cost is sufficiently close to lower bound.

The MP1 itself is decomposed into a master problem (MP2) and operation sub-problems (SP2). MP2, which is an integer problem, is solved to generate a trial solution for maintenance schedule decision variables. This master problem is a relaxation of the original problem as it contains only a subset of constraints. Its optimal value is a lower bound on the optimal value of original problem.

The critical point in the MP1 decomposition is the revision of MP2 based on the solution of fuel dispatch sub-problem. Associated with the solution of the fuel dispatch sub-problem is a set of shadow prices which measure changes in system operating costs caused by marginal changes in the trial maintenance. These shadow prices are used to form a linear constraint written in terms of maintenance variables

x. This constraint, known as Benders cut, is returned to the maintenance problem which is modified and solved again to determine a new trial maintenance plan.

3.11 DETAILED SOLUTION PROCESS

The formulation of the initial master problem, fuel dispatch sub-problems and the network sub-problem are given here. The revised master problem is the initial master problem in which Benders cuts are added to its constraints. The initial master problem is an integer program, but the network and fuel dispatch sub-problems are linear programs.

3.11.1 Initial Master Problem (MP2). It is the same as (3.16). Here, in order to proceed with a quick solution, we do not include emission constraints. Optimization of MP2 may be seen as a discrete optimization with decision variables x_{it}. Although, Branch and bound, dynamic programming or any other integer programming can solve this optimization, we use the branch and bound method to solve MP2.

3.11.2 Fuel Dispatch Sub-Problem (SP2). The fuel cost, w, depends on the utilization of available units to satisfy load and limited fuel constraints. The fuel dispatch sub-problem is formulated in (3.25).

$$Min \quad w = \sum_i \sum_m \sum_t c_{mt} \phi_{imt}$$

$$S.T. \qquad \sum_m \phi_{imt} \le K_i x_{it}^n \qquad \text{(dual variable is } \zeta\text{)}$$

$$\sum_i \phi_{imt} \le W_{mt}$$

$$\sum_i \sum_{t \in \text{month } j} \phi_{imt} \le W_{mj}$$

$$\sum_i \sum_{t \in year\ v} \phi_{imt} \le W_{mv} \qquad (3.25)$$

$$E\{g_{it}^j\} \le \sum_m \frac{\phi_{imt}}{h_{im}}$$

where x_{it}^n is the solution of MP2 at iteration n and $E\{g_{it}^n\}$ is the solution of SP1at iteration j. Optimization (3.25) is an LP problem and the shadow price which would be added to MP2 can be generated as:

$$z \ge w^n + \sum_t \left\{ \sum_i \left(C_{it}(1-x) + \zeta_{it}^n K_i (x_{it}^n - x_{it}) \right) + F_t^{j-1} \right\} \qquad (3.26)$$

It is important to point out that each generating unit is allowed to burn multiple constrained fuels in this formulation. As a result, the coordination of multiple constrained fuels which compete with one another becomes an imperative step for solving this problem. The consumption of one constrained fuel impacts the consumption of some other constrained fuels. In this case, sub-problem SP2 accommodates the general fuel allocation problem.

3.11.3 ISO (Network) Sub-Problem (SP1).

Long-term scheduling must ensure that a sufficient reserve quantity exists to provide power supply security while minimizing the cost of operation. Before minimizing the operation cost, we check the feasibility of the sub-problem at the weekly peak load as follows:

$$F_t = Min\ E\left\{\sum_i r_{it}\right\}$$

$$S.T.\ \boldsymbol{Sf} + \boldsymbol{g} + \boldsymbol{r} = \boldsymbol{D(\varphi)}$$

$$\boldsymbol{g} \le \boldsymbol{\bar{g}(\varphi).x}^n \qquad (dual\ variable\ is\ \lambda) \tag{3.27}$$

$$\boldsymbol{r} \le \boldsymbol{D(\varphi)}$$

$$|\boldsymbol{f}| \le \boldsymbol{\bar{f}(\varphi)}$$

The sub-problems may be infeasible due to the fact that the reliability constraint cannot be met. If one or more of sub-problems are infeasible, then an infeasibility cut is generated. For each infeasible sub-problem resulting from the jth solution of the master problem, the infeasibility cut is:

$$E\left\{\sum_i r_{it}^j\right\} + \sum_i \lambda_{it}^j\, \bar{g}_i\, (x_{it}^n - x_{it}) \le \varepsilon \qquad \forall t \in infeasible \tag{3.28}$$

If sub-problems are feasible for all t, then the energy purchased for period t, denoted as F_t, will depend on the utilization of available generating units to satisfy load constraints in each time period subject to maintaining the reliability above a certain level. Thus, the minimization of energy purchased from outside in period t can be expressed as:

$$F_t = Min\ \chi_t\, \tau \int_{\sum_{i \le N_g} \bar{g}_i x_{it}^n}^{\infty} ELDC_{N_g}(g)dg \tag{3.29}$$

where N_g = number of units

τ = number of hours in week t

$\bar{g}_i$ = maximum capacity of unit i

$ELDC_{Ng}$ = equivalent load duration curve after N_g units have been loaded

The solution of sub-problem SP1 is not complicated as we already know which generators are available during period t, which allows us to schedule them in the merit order according to their operating costs and load. We utilize each unit to capacity, starting with the cheapest unit until the reliability constraint is satisfied. If there are not enough units available to meet the reliability requirement, we would consider purchasing energy from other providers.

The solution of (3.29) for all t yields a set of dual multipliers from which a cost cut is constructed. The cost cut to be added to MP1 is of the following form:

$$z \geq \sum_t \left\{ F_t^j + \sum_i \left(C_{it}(1 - x_{it}) + \pi_{it}^j \, \bar{g}_i (x_{it}^n - x_{it}) \right) + w^n \right\} \tag{3.30}$$

where F_t^j is the expected purchased energy in period t associated with the jth trial solution. The fuel cost, w^n, is the latest solution of the fuel dispatch sub-problem (SP2). This fuel cost can be set to zero initially. The optimal dual multipliers are given as:

$$\pi_{it} = \frac{\partial F_t}{\partial (g_{it} x_{it})} \tag{3.31}$$

The multiplier π_{it}^j may be interpreted as expected marginal costs associated with 1 MW decrease in power generation, given the jth trial maintenance schedule. The cost cut (3.30) tends to increase the lower bounds obtained from successive maintenance sub-problem solutions.

Figure 3.5 demonstrates the method to find dual multipliers for generating unit i. Equivalent Load Duration Curve (ELDC) is derived based on the cumulant method [Str80]. A numerical example for calculating dual multipliers using the cumulant method is given in Appendix B. Cumulants exhibit highly desirable characteristics for our analysis: first, individual unit cumulants are calculated and used repeatedly throughout the analysis; second, the derivative term in (3.31) is determined using the Gram-Charlier approximation.

ELDC is evaluated at L_i after convolving all available units (which includes unit i operating in its installed capacity). Then, unit i is subtracted from the system load and ELDC is evaluated at L_{i-1}. If we assume that ELDC is linear in the region of interest, the dual multiplier of the i-th unit is approximated as:

$$\pi_{it} = \chi_t \tau \frac{(ELDC_{i-1}(L_{i-1})) + ELDC_i(L_i))}{2} \tag{3.32}$$

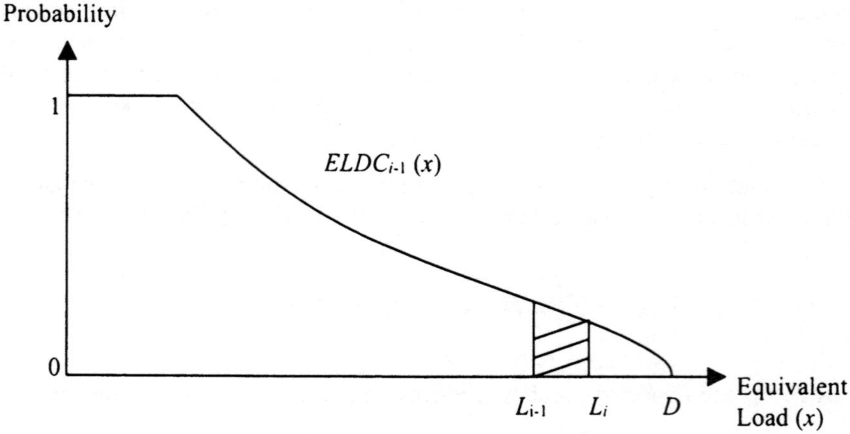

Figure 3.5 Equivalent Load Curve

3.11.4 Revised Master Problem (MP1). Network sub-problems (SP1) yield a set of dual multipliers from which a cut is constructed in each iteration. This cut is added to MP1 to revise the previous MP1. Emission constraints are included in MP1 to ensure that the emission level is below the allowable limit while minimizing the cost of operation.

$$Min \quad z$$

$$S.T. \quad z \ge \sum_{t} \sum_{i,k} \left\{ C_{it}(1 - x_{it}) \right\}$$

$$x_{it} = 1 \qquad for \; t \le e_i \; or \; t \ge l_i + d_i$$

$$x_{it} = 0 \qquad for \; s_i \le t \le s_i + d_i$$

$$x_{it} \in \{0,1\} \qquad for \; e_i \le t \le l_i$$

1. crew availability   *3. seasonal limitations*

2. resources availability   *4. desirable schedule*

• *emission constraints :*

$$\sum_{t} \sum_{i} H_i(E\{g_{it}\}) x_{it} \le EMS$$

$$\sum_{t} \sum_{i \in A} H_i(E\{g_{it}\}) x_{it} \le EMA$$

• *the cut from SP2 :*

$$z \geq w^n + \sum_t \left\{ \sum_i \left(C_{it}(1-x) + \zeta_{it}^n \, K_i(x_{it}^n - x_{it}) \right) + F_t^{j-1} \right\}$$

• *if SP1 is feasible for every t, then cut from SP1 :*

$$z \geq \sum_t \left\{ F_t^j + \sum_i \left(C_{it}(1-x_{it}) + \pi_{it}^j \, \overline{g}_i(x_{it}^n - x_{it}) \right) + w^n \right\}$$

• *if one or more sub - problems are infeasible, then the cut from SP1*

$$E\left\{ \sum_i r_{it}^j \right\} + \sum_i \lambda_{it}^j \, \overline{g}_i \, (x_{it}^n - x_{it}) \leq \varepsilon \qquad \forall t \in infeasible$$

$$x_{it} \in \{0,1\} \tag{3.33}$$

where: n is the index of inner iteration,

π^j, λ^j are multiplier vectors of jth outer iteration.

The proposed method is depicted in Figure 3.6.

The important feature of the Bender decomposition is the availability of upper and lower bounds to the optimal solution at each iteration. These bounds can be used convergence criteria. The convergence criterion for the inner loop in Figure 3.6 is:

$$\frac{2\left(z^n - z^{n-1}\right)}{\left(z^n + z^{n-1}\right)} \leq \Delta_1 \tag{3.34}$$

The convergence criterion for the outer loop in Figure 3.6 is:

$$\frac{2\left(z^j - z^{j-1}\right)}{\left(z^j + z^{j-1}\right)} \leq \Delta_2 \tag{3.35}$$

where z^n is the value of z at iteration n in the inner loop and z^j is the value of z at iteration j in the outer loop.

There is a trade-off between Δ and number of iterations. The smaller the Δ the larger the number of iterations will be. For some cases the problem may not converge if Δ is too small. In this book, setting Δ to 1% is good enough.

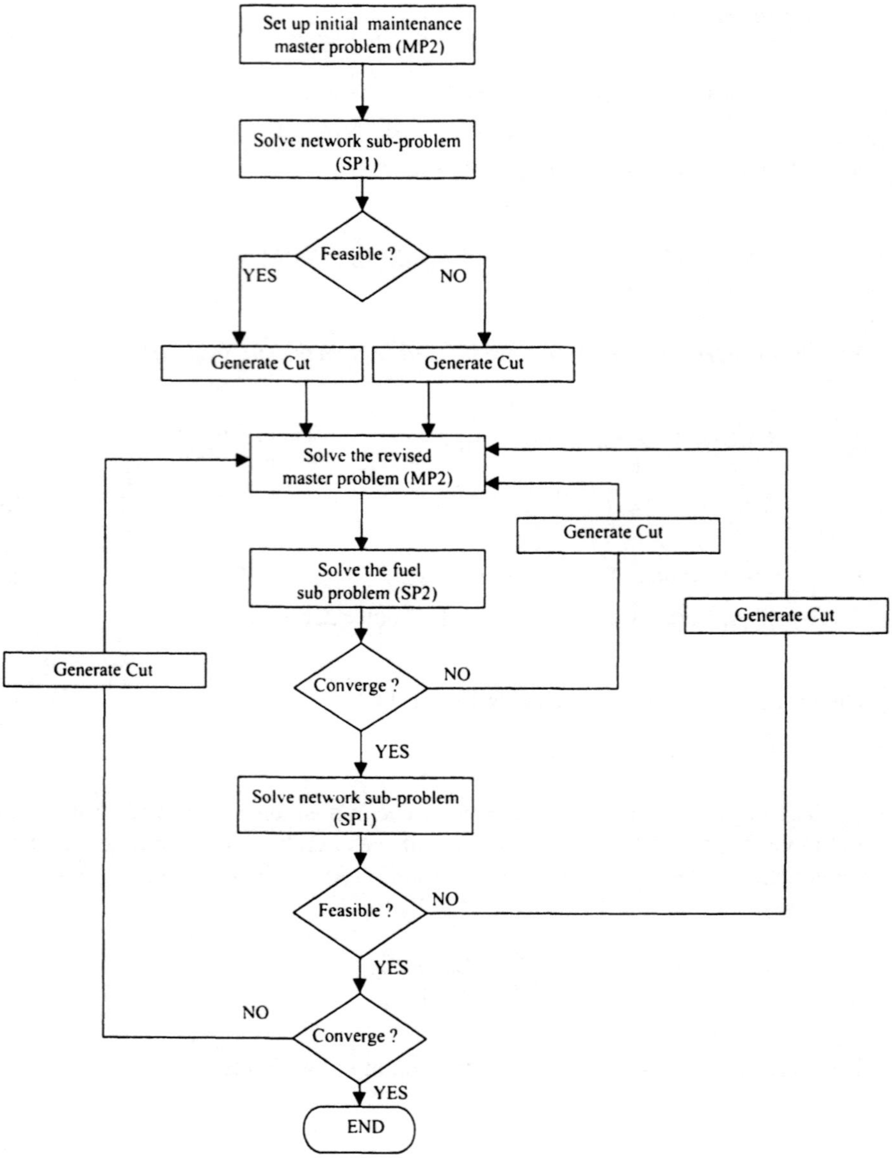

Figure 3.6 Double Decomposition Algorithm

3.12 CASE STUDIES

We apply IEEE-RTS discussed earlier in this chapter to this section. The emission coefficients are the same as the corresponding cost curve coefficients and the emission cap is the same as the peak load, with all values multiplied by a conversion factor of 0.8. The average fuel price is given in Table 3.23. The cost of per MWh of energy purchased from other providers is 49.00 $/MWh.

Table 3.23 Fuel Price

Fuel	Unit Size (MW)	Total Weekly Fuel Limit (MBtu)	Fuel Price ($/MBtu)
Oil #2	20	69,400	3.0
Oil #6	12 , 100 and 197	198,675	2.3
Coal	76 , 155 and 350	210,736	1.2

The results of the following test cases are included to show the effect of network, emission and fuel limits on maintenance.

- Case 3: Fuel limits are imposed on Case 2 in Section 3.8.
- Case 4: Emission constraints are imposed on Case 3.

Table 3.24 shows the comparative cost figures for total cost of generating units. The total cost when emission limits are not considered is $36.98856 million. Coal units 1 and 2 are cheap units and are loaded to their maximum at all times (base load units). As we impose emission requirements on the scheduling problem, a solution which satisfies emission requirements is reached. However, the total cost increases to $ 37.19351 million as compared to $36.98856 million when emission was not considered. In Case 4, generation of coal units 1 and 2 decreases from maximum at several periods to satisfy emission requirements and more oil units are committed to satisfy power demand.

Table 3.24 Maintenance and Production Cost of Generating Units (12-week horizon)

Case	Cost (10^6 $)
0	36.62694
1	36.67142
2	36.69147
3	36.98856
4	37.19351

Table 3.25 shows the status of different units in all four cases. In Case 3, fuel limits are imposed on the problem which have limited the output of all units.

In this table, Unit 3, which was always on-line in Cases 0 and 1, is forced to be down in weeks 22, 23. Unit 5 is on-line more often to compensate for unit 3 that has to be down. The maintenance schedule of unit 1 is also shifted to weeks 22-23. Case 4 represents the most expensive case for maintenance as cheaper units (e.g., unit 1) are turned off and more expensive units (e.g., unit 5) are turned on to satisfy the constraints.

Table 3.25 Generating Unit Schedule (12-week horizon)

Case	Unit	Weeks on Maintenance	Weeks On	Weeks Off
0	1	24,25	18-23,26-29	-
	2	27,28	18-26,29	-
	3	20,21	18,19,22-29	-
	4	18,19	20,21,24,25,27,28	22,23,26,29
	5	22,23	-	18-21,24-29
1	1	23,24	18-22,25-29	-
	2	28,29	18-27	-
	3	20,21	18,19,22-29	-
	4	18,19	21-24,28,29	20,25-27
	5	27,28	23,24	18-22,25,26,29
2	1	23,24	18-22,25-29	-
	2	28,29	18-27	-
	3	18,19	20-29	-
	4	20,21	18,19,23,24,28,29	22,25-29
	5	27,28	20-24	18-19,25,26,29
3	1	22,23	18-21,24-29	-
	2	28,29	18-27	-
	3	18,19	20,21,24-29	22,23
	4	20,21	18,19,22-24,28,29	25-27
	5	27,28	21-24	18-20,25,26,29
4	1	26,27	19-25,28,29	18
	2	28,29	18-20,22-27	21
	3	18,19	20,21,24-29	22,23
	4	20,21	18,19,22-24,28,29	25-27
	5	27,28	18,21-24	19,20,25,26,29

Next, the test system is applied to a year-long horizon of 52 weeks. Tables 3.26 shows the maintenance schedule for generating units, respectively. Since units are not forced on maintenance within 12 weeks, the maintenance cost is reduced as in Table 3.27 and a better distribution of risk is achieved as that in Figure 3.7. In this case, unit maintenance schedules have more flexibility to shift to lower peak loads, i.e., maintenance of unit 1 is shifted from weeks 26-27 to weeks 35-36. Also units 2, 4 and 5 are shifted from weeks 28-29 to weeks 37-38, weeks 20-21 to 10-11, and weeks 27-28 to 39-40, respectively.

Table 3.26 Generating Unit Maintenance in Case 4 (52-week horizon)

Unit	Weeks on Maintenance
1	35,36
2	37,38
3	18,19
4	10,11
5	39,40

Table 3.27 Total Cost of Generating Unit

Case	Cost ($\$10^6$)
12-week	37.19351
52-week	37.07934

Figure 3.8 shows the effect of maintenance scheduling on system reliability (***E***(*r*)). As we impose more constraints on the problem, the expected energy not served is increased which in turn increases the expected energy to be purchased from outside resources. From Figure 3.8, it can be observed that imported energy in week 22 of Case 3 is larger than that of Case 4. This is because of unit 1 of Case 3 which has been forced to be on maintenance in week 22 to minimize the total cost over the 12-week horizon. The benefit of considering unit 1 of Case 3 for maintenance in week 22 can be seen in weeks 25-29 in which the imported energy of Case 3 is much lower than that of Case 4.

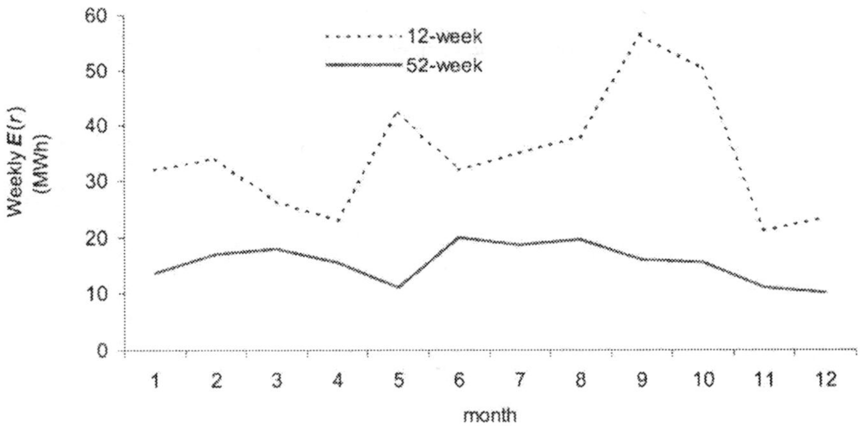

Figure 3.7 Effect of Maintenance Window on System Reliability

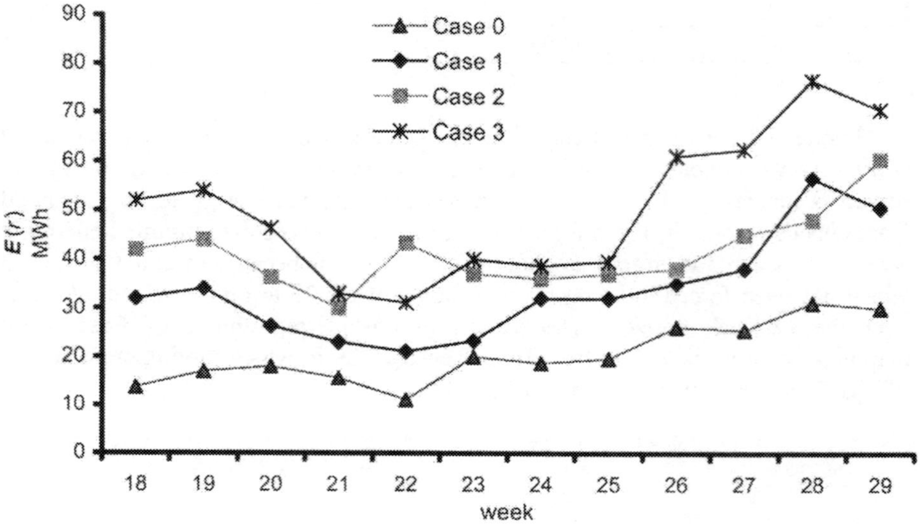

Figure 3.8 Scheduling Method Effects on System Reliability

CHAPTER IV

SHORT-TERM GENERATION SCHEDULING

4.1 INTRODUCTION

This chapter discusses the short-term generation scheduling (unit commitment) which will be used in the following chapter for the coordination between long-term and short-term generation maintenance scheduling.

Traditionally, in a vertically integrated utility environment, a utility system operator, who had the knowledge of system components, constraints and operating costs of generating units, was the one who made decisions on generating unit commitment for minimizing the utility's generation cost. In a restructured market-based system, committing generating units renders additional challenges in competition, market uncertainties, gaming and price forecast. Market Clearing Prices (MCPs), the price at which the generation meets the demand, are not known in advance to let GENCOs commit their units economically. Generating units are subject to charges when they are started up and incur additional costs when they are operated at some minimum level while waiting to be dispatched by GENCO operators. When generating units are dispatched, GENCOs recover their costs based on MCPs. Even though a GENCO may predict the hours that it will operate its units at a certain MCP to recover its costs, it is difficult for GENCOs to predict how many hours their generating units will actually be dispatched. This fact results in uncertainty on GENCOs' revenues.

Market structures can differ according to their participants, the amount of information that participants share with the ISO, and the role of ISO in facilitating or controlling the individual markets. The basic difference regarding the amount of control given to the ISO for generating unit commitment is presented as follows.

1) **Security Constrained Unit Commitment (SCUC)** -- The ISO will commit generating units in much the same way as system operators did in the vertically integrated structure. The ISO uses the bid-in costs submitted for each generating unit (multi-part bids that reflect start-up costs, minimum generation costs, running costs, etc.) to minimize the cost of operation, determine which units will be dispatched in how many hours and calculate the corresponding MCPs while the system security is retained. The objective in this case is to:

$$\textit{Minimize} \quad \textit{Bid-in Cost} \qquad \text{(I)}$$

   SCUC is used in Pennsylvania-New Jersey-Maryland (PJM) interconnection, New York Market, and U.K. Power Pool.

2) **Price Based Unit Commitment (PBUC)** – GENCOs' decision to commit generating units is associated with financial risks. Each energy supplier would be responsible for its own decision on what and how to bid on energy to supply

load and ancillary services markets. In this approach, bidders bear all the risks of their decisions for committing their units and bidding strategies. GENCOs run their own price-based unit commitment where the objective is to maximize individual GENCO's profit, regardless of the system wide profit. GENCOs submit single-part bids to the ISO for minimizing the risk of not knowing the number of hours that they would be dispatched or MCPs that would be paid on during dispatched hours. The ISO uses GENCOs' single-part bids, aggregates them and determines the MCP. Here, the ISO's objective will be to maintain the system security, while the GENCO's objective will be to maximize its own profit:

$$Maximize \quad Profit = Revenue - Cost$$

which is equivalent to:

$$Minimize \quad (-Profit) = Cost - Revenue \qquad \text{(II)}$$

This approach is used in New England Power Pool, California Market, New Zealand, and Australia.

It is easy to realize that (I) is not equivalent to (II) even if GENCOs use the marginal cost for generating units. In this chapter, we first discuss SCUC. Then we provide a PBUC model for trading energy and ancillary services.

4.2 SECURITY-CONSTRAINED UNIT COMMITMENT

The ISO runs SCUC and schedules generating units for minimizing bid-in costs submitted by GENCOs and satisfying the prevailing constraints such as load balance, spinning reserve, ramp rate limits, fuel constraints, multiple emission requirements as well as minimum up and down time limits over a set of time periods. The ISO also obtains information on transmission capacity and availability from TRANSCOs via the Open Access Same Time Information System (OASIS). By applying SCUC, the ISO schedule generating units for supplying load demands while maintaining transmission flows and bus voltages within their permissible limits. The supply of reactive power (ancillary service) by the ISO into a system would improve the quality of service and the economic operation of the system. This goal can be accomplished by committing and dispatching reactive power generation, utilizing capacitors, adjusting tap changing transformers.

In a well-balanced system, there is a likelihood that we would not need to run SCUC, and a traditional unit commitment will schedule generating units and satisfy loads economically. However, if most of the committed units (cheaper units) are located in one zone of the system, there is a chance that the flow of power out of that zone would overload transmission lines and drop bus voltages at remote locations. In such cases, available generating units, phase shifters and tap transformers could be utilized by the ISO to alleviate transmission overflows and improve the voltage profile in all zones for a given set of committed units.

As the system becomes more congested with loop flows and wheeled power flows, either control devices would have a more difficult time in adjusting transmission flows for alleviating violations or the adjusted flows would deviate significantly from economically dispatched flows (preferred schedules). In such cases, SCUC plays a major role in system operation and maintaining its security as NERC has mandated for control areas to maintain sufficient ancillary services under n-1 contingencies.

4.2.1 SCUC Problem Formulation. SCUC follows the traditional unit commitment formulation in which the system load is calculated by the ISO. SCUC is formulated as an optimization problem which minimizes the operating cost.

$$PC = \sum_i \sum_\tau \left[I_{i\tau} F_i(g_{i\tau}) + S_{i\tau} \right] \tag{4.0}$$

In a GENCO, the production cost $F_i(g_{i\tau})$ is calculated as the product of the heat rate (MBTU/hr) and the unit's fuel cost ($/MBTU). The start up cost of a unit depends on the length of time that the unit has been off and is defined as:

$$S_{i\tau} = I_{i\tau}[1 - I_{i(\tau-1)}]\left[\alpha_i + \beta_i\left(1 - exp\frac{-Z_{i\tau}^{off}}{\Theta_i}\right)\right]$$

The constraints of the optimization problem are:

System real power balance

$$\sum_i I_{i\tau} g_{i\tau} = P_D(\tau) \qquad \tau = 1, \ldots, t \tag{4.1}$$

Spinning reserve is the unloaded synchronized generation that can ramp up in 10 minutes. Non-spinning reserve is a generating capacity, not synchronized to the grid, that can ramp up in 10 minutes.

System spinning reserve requirements

$$\sum_i r_{si}(\tau) I_{i\tau} \geq R_s(\tau) \qquad \tau = 1, \ldots, t \tag{4.2}$$

where, $r_{si}(\tau) = min\{10 * MSR_i, \bar{g}_i - g_{i\tau}\}$. The spinning reserve requirement, $R_s(\tau)$, is typically defined as a base component plus a fraction of the load requirement and a fraction of the high operating limit of the largest on-line unit.

System operating reserve requirements

$$\sum_i r_{oi}(\tau)I_{i\tau} \geq R_o(\tau) \qquad \tau = 1,\ldots, t \tag{4.3}$$

where, $r_{oi}(\tau) = \begin{cases} q_i, & \text{if unit } i \text{ is OFF} \\ r_{si}(\tau), & \text{if unit } i \text{ is ON} \end{cases}$

Interruptible loads are usually added to the operating reserve capacity of units (left hand side of Eq. (4.3)). The operating reserve requirement, $R_o(\tau)$, is commonly defined similar to $R_s(\tau)$ as a function of a base component plus a fraction of the load requirement and a fraction of the high operating limit of the largest on-line unit.

Unit generation limits

$$\underline{g}_i \leq g_{i\tau} \leq \overline{g}_i \qquad \tau = 1,\ldots, t \qquad \forall i \tag{4.4}$$

Thermal unit minimum start up/down times

$$(Z^{on}_{i(\tau-1)} - T^{on}_i)*(I_{i(\tau-1)} - I_{i\tau}) \geq 0 \tag{4.5}$$

$$(Z^{off}_{i(\tau-1)} - T^{off}_i)*(I_{i(\tau-1)} - I_{i\tau}) \geq 0 \tag{4.6}$$

Ramp rate limits

$$g_{i\tau} - g_{i(\tau-1)} \leq UR_i \quad \text{as unit } i \text{ ramps up} \tag{4.7}$$

$$g_{i(\tau-1)} - g_{i\tau} \leq DR_i \quad \text{as unit } i \text{ ramps down} \tag{4.8}$$

Fuel constraints

$$FL \leq \sum_\tau \sum_{i \in A} f_{ci} * [I_{i\tau}F_i(g_{i\tau}) + S_{i\tau}] \leq FU \tag{4.9}$$

where $i \in A$ corresponds to all fuel-constrained units in group A. The first item $f_{ci}I_{i\tau}F_i(g_{i\tau})$ is operating fuel and $f_{ci}S_{i\tau}$ is additional fuel for start up.

System emission limit

$$\sum_\tau \sum_i H_i(g_{i\tau})I_{i\tau} \leq EMS_t \tag{4.10}$$

where several emission types (e.g. *SO2*, *NOx*) are considered.

Area emission limit

$$\sum_{\tau}\sum_{i\in B} H_i(g_{i\tau})I_{i\tau} \le EMA_t \tag{4.11}$$

where $i \in B$ corresponds to all units in the constrained emission area B where several emission types can be considered.

Transmission flow limits of line k

$$\underline{f}_k \le f_{k\tau} = f(g_{i\tau},\delta_\tau) \le \overline{f}_k \qquad for \quad \tau \in t \tag{4.12}$$

where δ_τ is the phase shifter control vector

Reactive power operating reserve requirement

$$\sum_{i} I_{i\tau}\overline{Q}_i \ge Q_D(\tau) \tag{4.13}$$

Reactive power generation limits and load bus balance

$$\underline{Q}_i I_{i\tau} \le Q_{i\tau} = F_1(V) \le \overline{Q}_i I_{i\tau} \tag{4.14}$$

$$Q_D(\tau) = F_2(V) \tag{4.15}$$

System voltage and transformer tap limits

$$\underline{V} \le V \le \overline{V} \tag{4.16}$$

$$\underline{T} \le T \le \overline{T} \tag{4.17}$$

In the above formulation, (4.1-4.12) are real power generation constraints. The constraint (4.12) is the ac transmission flow limit. The transmission flow depends on unit generation and phase shifter controls. In the SCUC solution, if all transmission constraints are relaxed in the Lagrangian objective function, the problem will become vastly complicated and impossible to solve with a large number of Lagrangian multipliers. If we exclude the network constrains (4.12-4.17) from the above set of equations, the rest represents the traditional formulation for unit commitment which has been solved before. So a possible alternative for the inclusion of network constraints in unit commitment is to introduce a Benders decomposition which would decompose SCUC into a master problem and a subproblem, and applies an iterative approach to solve the problem. The solution procedure is shown in Figure 4.1. The other advantage of this modularized approach is that it utilizes the existing unit commitment package in the master problem and hence reduces the amount of software modification required for establishing SCUC. In the following, we discuss the solution of SCUC.

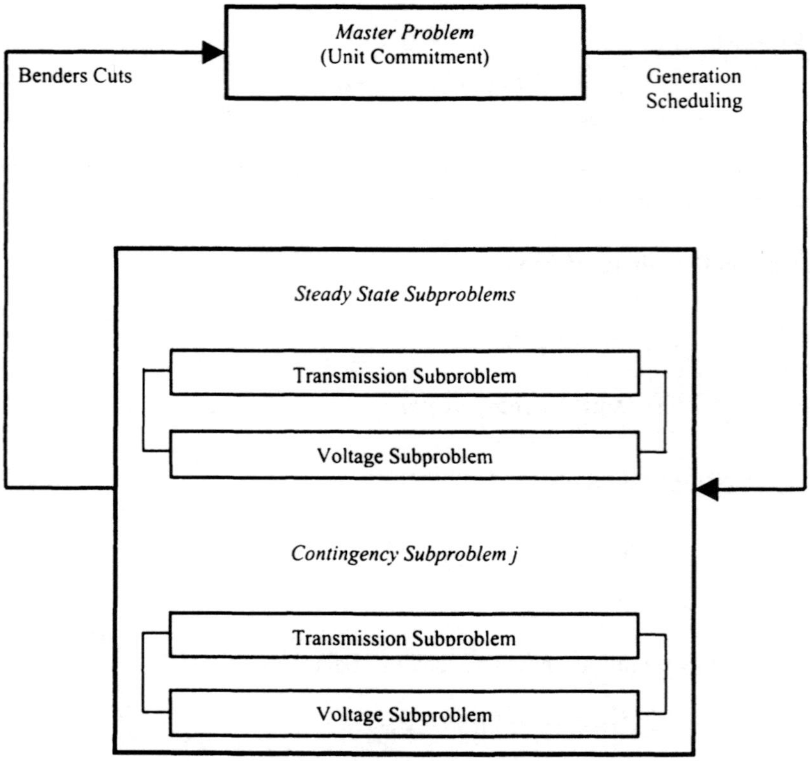

Figure 4.1 SCUC Problem with Transmission and Voltage Constraints

The master problem in this case solves the traditional unit commitment, by augmented Lagrangian relaxation, with all the prevailing constraints except the network constraints. Given the unit commitment schedule, the subproblem minimizes network violations. Since we apply a linearized network in this study, the subproblem would be decoupled into two smaller subproblems in each case, corresponding to transmission and voltage constraints. The transmission subproblem minimizes transmission flows for steady state and n-1 contingencies by adjusting unit generations and phase shifters. The reactive subproblem examines and removes voltage constraints by adjusting reactive power generation and tap changers. Then, Benders cuts will be generated if violations in the subproblems persist. By including Benders cuts in the master problem, unit commitment is solved again to provide a minimum cost generation schedule while, in this case, satisfying transmission security and voltage constraints.

The SCUC solution becomes more manageable and faster based on this approach since the decomposed problem is easier to solve, requires less complicated and smaller computing capabilities, and applies minor revisions to the traditional unit commitment package for considering network constraints.

4.2.2 Analysis of Constraints. Since we would primarily be interested in screening transmission violations in the subproblem, we use a linear model (dc load flow) to expedite the process. We use linear sensitivity factors to formulate transmission constraints that are given by:

1. Steady state constraints (no contingencies)

$$\underline{f}_k \le f_{k\tau} = \sum_i A^i_{km} (g_{i\tau} + P_{\phi i\tau} - d_{i\tau}) \le \overline{f}_k \tag{4.18}$$

2. Contingency constraints

$$\begin{aligned} &\underline{f}_k \le f_{k\tau} = \sum_i E^i_{km}(j)(g_{i\tau} + P_{\phi i\tau} - d_{i\tau}) \le \overline{f}_k \\ &where, \\ &j = 1, 2, ..., nc \end{aligned} \tag{4.19}$$

Derivation of transmission terms A^i_{km} and $E^i_{km}(j)$ is discussed in [Woo96]. (4.12-4.17) are binding constraints for unit commitment state $\boldsymbol{I}$. (4.16-4.17) represent bus voltage and transformer tap limits which are related to the unit commitment state $\boldsymbol{I}$ by (4.13 and 4.14).

We assume $\boldsymbol{V^0}$ is the system voltage vector based on the initial unit commitment state. If $\boldsymbol{V^0}$ is infeasible, additional adjustments to the generation and/or distribution of reactive power may be necessary in order to shift the system voltage from $\boldsymbol{V^0}$ to a desired value V^*. We neglect the effect of reactive power adjustment on bus voltage angles. (4.14-4.15) are linearized in the vicinity of the initial operating point $\boldsymbol{V^0}$ and the effect of tap-changing transformer on voltages is considered in the linearized form as follows (see Chapter 2):

1. Steady state constraints (no contingencies)

$$\Delta \underline{\boldsymbol{Q}}_T + \underline{\boldsymbol{Q}}b_i \le \boldsymbol{Q}\, b_i^0 + \boldsymbol{J}_1^{\prime\prime s}\, \Delta V \le \overline{\boldsymbol{Q}}b_i + \Delta \overline{\boldsymbol{Q}}_T \tag{4.20}$$

$$\boldsymbol{J}_2^{\prime\prime s}\, \Delta V = \boldsymbol{0} \tag{4.21}$$

$$\Delta \underline{V} \le \Delta V \le \Delta \overline{V} \tag{4.22}$$

2. Contingency constraints

$$\Delta \underline{\boldsymbol{Q}}_T + \underline{\boldsymbol{Q}}b_i \le \boldsymbol{Q}\, b_i^0 + \boldsymbol{J}_1^{\prime\prime c}\, \Delta V \le \overline{\boldsymbol{Q}}b_i + \Delta \overline{\boldsymbol{Q}}_T \tag{4.23}$$

$$\boldsymbol{J}_2^{\prime\prime c}\, \Delta V = \boldsymbol{0} \tag{4.24}$$

$$\Delta \underline{V} \le \Delta V \le \Delta \overline{V} \tag{4.25}$$

Where $\Delta V = V^* - V^0$ is the incremental system voltage corresponding to ΔQ_i . $J_1''^S$ and $J_1''^C$ are the modified Jacobian matrices for buses connected to generators and transformers in steady state and contingency cases respectively. Also, $J_1''^S$ and $J_1''^C$ are the modified Jacobian matrices for load buses in steady state and contingency cases respectively.

4.3 APPLICATION OF DECOMPOSITION

The two sets of inequalities in SCUC represent real power constraints given in (4.1-4.11), and transmission security and voltage constraints given in (4.12-4.17) which are rewritten as (4.18-4.25). By representing the above transmission security and voltage constraints in the unit commitment formulation, the SCUC problem can be generalized as follows:

$$\begin{aligned} &Min\ \boldsymbol{ux} \\ &St.\ \ \boldsymbol{Ax} \geq \boldsymbol{b} \\ &\qquad \boldsymbol{Ex} + \boldsymbol{Fy} \geq \boldsymbol{h} \end{aligned} \tag{4.26}$$

where $\boldsymbol{x}$ represents the unit commitment state (master problem), and $\boldsymbol{y}$ represents the penalty variable for satisfying constraints in subproblems. The formulation (4.26) is a standard form of Benders decomposition which is solved as follows:

(i) In the master problem, the unit commitment state $\boldsymbol{x}$ is calculated as follows:

$$\begin{aligned} &Min\ \ \boldsymbol{ux} \\ &St.\ \ \boldsymbol{Ax} \geq \boldsymbol{b} \\ &\qquad w(\boldsymbol{x}) \leq 0 \end{aligned} \tag{4.27}$$

where $w(\boldsymbol{x})$ is the cut added to the master problem to represent the infeasibility of the unit commitment state $\boldsymbol{x}$ in terms of transmission security and voltage constrains. The formulation of $w(\boldsymbol{x})$ will be discussed later. A typical unit commitment approach (augmented Lagrangian relaxation, dynamic programming, branch and bound, artificial intelligence, etc.) may be used for solving (4.27).

(ii) Given $\hat{\boldsymbol{x}}$, the subproblem is formulated as follows:

$$\begin{aligned} &Min\ \ w(\hat{\boldsymbol{x}}) = \boldsymbol{dy} \\ &St.\ \ \boldsymbol{Fy} \geq \boldsymbol{h} - \boldsymbol{E}\hat{\boldsymbol{x}} \end{aligned} \tag{4.28}$$

If the objective function $w(\hat{\boldsymbol{x}})$ larger than zero, then there would be a violation in the subproblem. We produce Benders cuts $w(\boldsymbol{x}) \leq 0$ based on a linear approximation in which the coefficients of linear approximation are the simplex multipliers π_i associated with constraints in (4.26). The linear form of the Benders cut is

$$w(\boldsymbol{x}) = w(\hat{\boldsymbol{x}}) + \boldsymbol{\pi}(\boldsymbol{x} - \hat{\boldsymbol{x}}) \leq 0 \tag{4.29}$$

where $w(\hat{\boldsymbol{x}})$ is the optimal solution of (4.28)

$\hat{\boldsymbol{x}}$ is the solution for the master problem

$\boldsymbol{\pi}$ is the simplex multiplier vector

$\pi_i = \dfrac{\partial w}{\partial x_i}$ is the simplex multiplier in linear programming

4.4 SCUC SOLUTION

Based on the earlier discussion, we solve the SCUC problem as follow:

Stage a: Solve unit commitment problem (master problem, 4.1-4.11) without transmission security and voltage constraints.

Stage b: With unit commitment schedule in stage a, test transmission security and voltage violations respectively in the subproblem (4.12-4.17). Benders cuts are produced for unit commitment rescheduling if any violation is detected at this stage.

4.4.1 Master Problem. The master problem solves unit commitment without transmission security and voltage constraints. At present, one of the potential approaches for solving unit commitment is the Lagrangian Relaxation (LR) method. The basic idea of LR is to relax system constraints in the objective function by using Lagrangian multipliers. The relaxed problem is then decomposed into N subproblems for each generating unit. The dynamic programming process is used to search the optimal commitment for single units. The Lagrangian multipliers are updated based on violations of system constraints. The convergence criterion is satisfied if the duality gap in Lagrangian relaxation is within a prescribed limit.

The quality of the final LR solution depends on the sensitivity of the commitment to Lagrangian multipliers. Slow and unsteady convergence of LR has always been a problem in finding the optimal solution as reported in most unit commitment solutions. Unless a proper modification of multipliers is ensured in every iteration, an unnecessary commitment of generating units may occur, which may result in higher production costs.

The difficulties in solving the unit commitment are often explained by the non-convexity of the optimization problem. The non-convexity problem may be overcome by augmented Lagrangian in which quadratic penalty terms associated with power demand are added to the objective function to improve the convexity of the problem. The augmented Lagrangian function for unit commitment is:

$$\Gamma(g_{i\tau}, I_{i\tau}, \lambda(\tau), \mu_s(\tau), \mu_0(\tau), \lambda_f, \lambda_s, \lambda_a)$$

$$= \sum_i \sum_\tau I_{i\tau} F_i(g_{i\tau}) - \sum_\tau \lambda(\tau) \sum_i I_{i\tau} g_{i\tau}$$

$$- \frac{\kappa}{2} \sum_\tau \left(\sum_i I_{i\tau} g_{i\tau} - P_D(\tau)\right)^2 - \sum_\tau \mu_s(\tau) \sum_i r_{si}(\tau) I_{i\tau}$$

$$- \sum_\tau \mu_o(\tau) \sum_i r_{oi}(\tau) I_{i\tau} + \lambda_f \sum_\tau \sum_{i \in F} f_{ci} * F_i(g_{i\tau})$$

$$+ \lambda_s \sum_\tau \sum_i H_i(g_{i\tau}) + \lambda_a \sum_\tau \sum_{i \in A} H_i(g_{i\tau}) I_{i\tau} \quad (4.30)$$

in which the terms that are independent of the decision variable have been dropped. κ is a positive penalty factor associated with the augmented term. The augmentation would cause the Lagrangian objective function to be continuously differentiable and strongly convex. Hence, oscillations and ill-conditioning in the unit commitment solution could be avoided.

Obviously, the quadratic term in the above equation is not separable and disrupts the decomposition of (4.30). However, by applying the decomposition and coordination technique, the non-separable quadratic penalty terms are linearized around the solution obtained from the previous iteration (see 4.31). Moreover, in order to improve the convergence property, we impose quadratic terms of decision variables in (4.31) to limit them from deviating too far from the last computed solution. These quadratic terms are separable, convex and differentiable with respect to $g_{i\tau}$. It should be emphasized that the added quadratic terms do not affect the optimality since they vanish at the optimal solution.

The optimization problem in $(k+1)^{th}$ iteration can be decoupled into N subproblems, each corresponding to the optimal unit commitment of individual units over the entire study period expressed as:

$$\Gamma(g_{i\tau}, I_{i\tau}, \lambda(\tau), \mu_s(\tau), \mu_0(\tau), \lambda_f, \lambda_s, \lambda_a)$$

$$= \sum_\tau \Big\{ F_i(g_{i\tau}) - \hat{\lambda}(\tau) g_{i\tau} - \hat{\mu}_s(\tau) r_{si}(\tau) - \hat{\mu}_o(\tau) r_{oi}(\tau)$$

$$- \kappa \cdot g_{i\tau} [\sum_{i=1} I_{i\tau}^k g_{i\tau}^k - P_D(\tau)] \quad (4.31)$$

$$+\hat{\lambda}_f f_{ci} * F_i(g_{i\tau}) + \hat{\lambda}_s H_i(g_{i\tau}) + \hat{\lambda}_a H_i(g_{i\tau})\Big\} I_{i\tau}$$

$$+\sum_{\tau}(S_{i\tau} + \kappa \cdot P_D(\tau) \cdot I_{i\tau}^k g_{i\tau}^k + \frac{1}{2\sigma}[I_{i\tau} g_{i\tau} - I_{i\tau}^k g_{i\tau}^k]^2)$$

subject to unit constraints (4.4-4.8).

At this stage, $g_{i\tau}$ is set equal to the unit power generation corresponding to the Lagrangian multiplier $\lambda(\tau)$. Consequently, integer variables $I_{i\tau}$, $\tau = 1,\ldots,n_\tau$ are the only unknown values. We adopt dynamic programming to find $I_{i\tau}$ for minimizing (4.31). The value of the penalty factor κ associated with the augmented terms is computed as the difference between the largest and smallest system lambdas divided by the difference between the load values in the corresponding two time intervals. For better convergence property, σ is taken as $1/\sigma > 2\kappa$.

Once the generating units are committed, economic dispatch will be solved as a continuous variable optimization process subject to a set of system (coupling) and unit constraints, e.g., system load balances (4.1), unit generation limits (4.4) and ramping limits (4.7-4.8). A piecewise linear fuel cost function is adopted in our economic dispatch, so we may use LP to minimize the objective function. There is a possibility that we encounter ramping violations at some hours, if we start the optimization process from $\tau = 1$, especially at those hours with maximum system demand or those with small difference between committed capacity and system demand. Therefore, we initiate the economic dispatch at the hour with maximum system demand. At that hour, the objective function will be minimized without ramp rate constraints. Then we consider ramping and proceed forward and backward to optimize the generation schedule at other time intervals. In some cases when a ramping violation is encountered, we re-initiate the scheduling process at the hour with smallest difference between the committed capacity and system demand. Then the scheduling process proceeds forward and backward to include other time intervals.

4.4.1.1 Effect of augmented terms. Here, the effect of augmented terms on the convergence of a typical unit commitment solution is shown in Figure 4.2. For the augmented solution, the amount of duality gap drops progressively with a proper convergence property for the solution. When augmented terms are deleted, the solution oscillates significantly as shown in Figure 4.2 and the duality gap never drops to that with augmented terms. Finally, without augmentation, the system production cost may be higher and even an infeasible unit commitment schedule may occur.

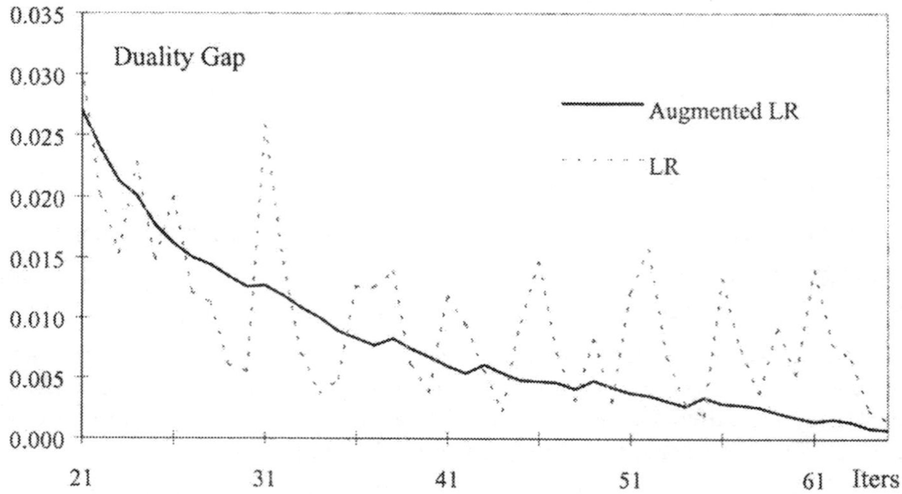

Figure 4.2 Duality Gaps vs. Number of Iterations in Unit Commitment

4.4.2 Subproblems Formulation.

The infeasibility of transmission security and voltage constraints is represented by adding penalty variables to transmission and voltage constraints. Penalty variables are interpreted as the amount of transmission power and voltage constraint violations associated with the unit commitment state $\hat{I}$. Therefore, we define the minimization of violations as the subproblem objective function $w(\hat{I})$ at each hour.

The infeasibility of voltage and transmission constraints is represented by adding penalty (slack) variables to voltage and transmission constraints, respectively. The penalty variables are interpreted as the amount of voltage and transmission violations associated with the base case unit commitment state $\hat{I}$. The matrix form of formulation is as follows:

$$w(\hat{I}) = Min \ \left\{ \ G^s + \sum_{j=1}^{nc} G^c(j) + \ F^s + \sum_{j=1}^{nc} F^c(j) \ \right\} \qquad j=1,2,...nc \tag{4.32}$$

$$St. \quad A \cdot g \ + \ F^s \ \leq \ f^s \tag{4.33}$$

$$J''^s \cdot \Delta V \ + G^s \leq \ g^s \tag{4.34}$$

$$E(j) \cdot g \ + \ F^c(j) \ \leq \ f^c(j) \qquad j=1,2,...nc \tag{4.35}$$

$$J''^c(j) \cdot \Delta V \ + G^c(j) \leq \ g^c(j) \qquad j=1,2,...nc \tag{4.36}$$

$$\underline{g} \leq g \leq \overline{g} \tag{4.37}$$

$$\Delta\underline{V} \leq \Delta V \leq \Delta\overline{V} \tag{4.38}$$

where (4.33-4.34) represent the steady state and (4.35-4.36) are for contingency cases. The objective function represents the minimization of penalty variables which may be multiplied by weighting factors in the objective function. $\boldsymbol{g}$ is the real power generation vector and ΔV is the increment voltage vector. Both are functions of unit commitment vector $\boldsymbol{I}$. The formulation (4.32-4.38) consists of several independent linear programs that could be decomposed into steady state and contingency subproblems.

4.4.2.1 Steady state subproblems. The formulation is as follows:

- **Transmission subproblems:**

$$w_S(\hat{\boldsymbol{I}}) = Min\{\ \boldsymbol{F}^S\ \} \tag{4.39}$$

$$St. \quad \boldsymbol{A}\cdot\boldsymbol{g} + \boldsymbol{F}^S \leq \boldsymbol{f}^s \tag{4.40}$$

$$\underline{\boldsymbol{g}} \leq \boldsymbol{g} \leq \overline{\boldsymbol{g}} \tag{4.41}$$

There are 24 steady state transmission subproblems that correspond to the 24 hour horizon.

- **Voltage subproblems:**

$$w_S(\hat{\boldsymbol{I}}) = Min\{\ \boldsymbol{G}^S\ \} \tag{4.42}$$

$$St. \quad \boldsymbol{J}''^s\cdot\Delta V + \boldsymbol{G}^S \leq \boldsymbol{g}^s \tag{4.43}$$

$$\Delta\underline{V} \leq \Delta V \leq \Delta\overline{V} \tag{4.44}$$

There are 24 steady state voltage subproblems that correspond to the 24 hour horizon.

4.4.2.2 Contingency subproblems. In each contingency case, the following subproblem is solved for each possible line outage j considered in that contingency case:

- **Transmission Subproblems:**

$$w_C(\hat{\boldsymbol{I}}, j) = Min\{\ \boldsymbol{F}^c(j)\ \} \qquad j=1,2,\ldots nc \tag{4.45}$$

$$St. \quad \boldsymbol{E}(j)\cdot\boldsymbol{g} + \boldsymbol{F}^c(j) \leq \boldsymbol{f}^c(j) \qquad j=1,2,\ldots nc \tag{4.46}$$

$$\underline{\boldsymbol{g}} \leq \boldsymbol{g} \leq \overline{\boldsymbol{g}} \tag{4.47}$$

There are $24 * nc$ transmission contingency subproblems that correspond to the 24 hours horizon.

- **Voltage subproblem:**

$$w_C(\hat{\boldsymbol{I}}, j) = Min\{\ \boldsymbol{G}^c(j)\ \} \qquad j=1,2,\ldots nc \tag{4.48}$$

$$St.\ \boldsymbol{J}''^{c}(j)\cdot\Delta\boldsymbol{V}+\boldsymbol{G}^{c}(j)\leq\boldsymbol{g}^{c}(j)\qquad j=1,2,\ldots nc \tag{4.49}$$

$$\Delta\underline{V}\leq\Delta V\leq\Delta\overline{V} \tag{4.50}$$

There are $24*nc$ voltage contingency subproblems that correspond to the 24 hours horizon.

4.4.3 Benders Cuts. If a violation is detected in the corresponding subproblem, the optimal values of $w_s(\hat{\boldsymbol{I}})$ or $w_c(\hat{\boldsymbol{I}},j)$ will be larger than zero. A linear approximation of the Benders cut is generated based on the subproblem results as:

$$w(\boldsymbol{I})=w(\hat{\boldsymbol{I}})+\boldsymbol{\pi}(\boldsymbol{I}-\hat{\boldsymbol{I}})\leq 0 \tag{4.51}$$

where $\boldsymbol{I}$ represents the unit commitment state, $\hat{\boldsymbol{I}}$ represents the base case unit commitment state and π_i is the simplex multiplier associated with the corresponding constraint in the subproblem.

4.4.4 Complications. For any contingency of nc possible single line outages in unit commitment, the following subproblems should be solved:

- 24 steady state transmission subproblems which may generate up to 24 Benders cuts.
- 24 steady state voltage subproblems which may generate up to 24 Benders cuts.
- $24*nc$ transmission contingency subproblems which may generate up to $24*nc$ Benders cuts.
- $24*nc$ voltage contingency subproblems which may generate up to $24*nc$ Benders cuts.

So, the total number of subproblems in each contingency case is $24*(1+1+nc+nc)$ which is $48*(1+nc)$. With such a large number of constraints added to the master problem, the problem becomes bigger and more complicated with a larger CPU time. However, the linear representation of these constraints minimizes the required CPU time. The added constraints are as follows:

$$w_s^i(\hat{\boldsymbol{I}})\ \leq\ 0.0\qquad i=1,2,\ldots 24 \tag{4.52}$$

$$w_c^i(\hat{\boldsymbol{I}},j)\ \leq\ 0.0\qquad i=1,2,\ldots 24\qquad j=1,2,\ldots nc \tag{4.53}$$

Figure 4.3 depicts the flowchart of SCUC.

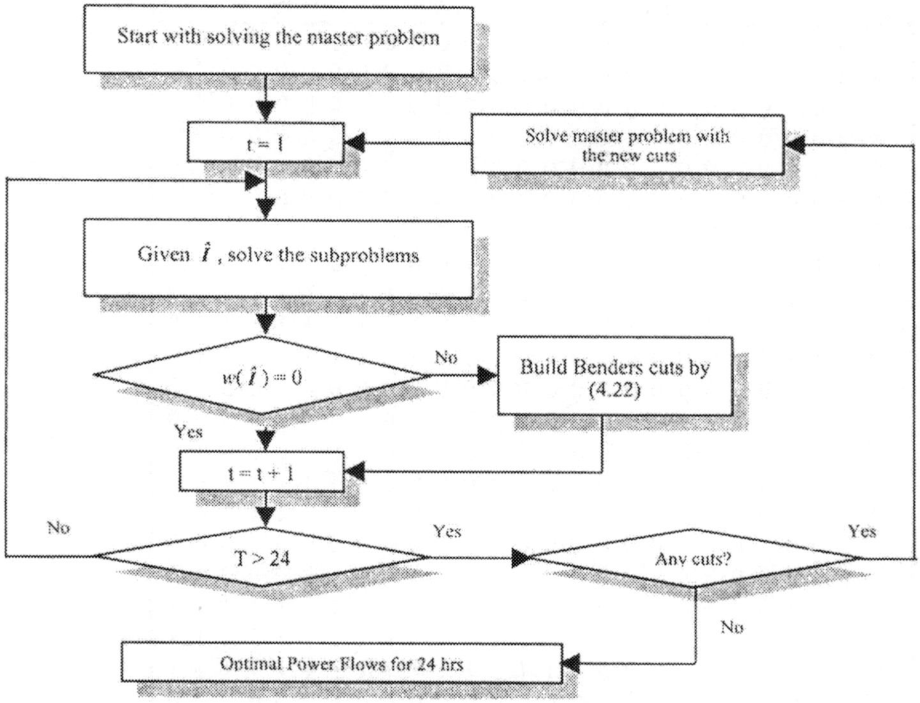

Figure 4.3 Flowchart of SCUC

4.5 CASE STUDIES

We provide studies for different cases to test the proposed SCUC. We assume loads are of constant power type, and use the IEEE-118 bus network with 36 generating units and 186-transmission lines (see Appendix E) to test the proposed approach. The one-line diagram of the IEEE-118 bus system is shown in Appendix E. Transmission lines have a 300 MW capacity. Lower and upper voltage limits are 0.97 and 1.06 p.u. respectively. The hourly load and spinning reserve distributions over a 24 hours horizon and characteristics of generators are given in Appendix E. The convergence criterion is given by a duality gap less than or equal to 0.00015.

4.5.1 Base Case. In the base case, no transmission or voltage constraints are considered (i.e., short-term unit commitment) which has a total production cost of $889,871.7 with an execution time of 25 sec. Units 28-36 in Table 4.1 constitute base load units which supply most of the system load. Intermediate and peaking units are also committed to satisfy the prevailing constraints.

Table 4.1 Unit Commitment without Network Constraints

No.	Hours (0-24)																							
1	0	0	0	0	0	0	0	0	0	0	0	0	0	0	0	0	0	0	1	0	0	1	0	0	0
2	0	0	0	0	0	0	0	0	0	0	0	0	0	0	0	0	0	0	0	0	0	0	0	0	0
3	0	0	0	0	0	0	0	0	0	0	0	0	0	0	0	0	0	0	0	0	0	0	0	0	0
4	0	0	0	0	0	0	0	0	0	0	0	0	0	0	0	0	0	0	0	0	0	0	0	0	0
5	0	0	0	0	0	0	0	0	0	0	0	0	0	0	0	0	0	0	0	0	0	0	0	0	0
6	0	0	0	0	0	0	0	0	0	0	0	0	0	0	0	0	0	0	0	0	0	0	0	0	0
7	0	0	0	0	0	0	0	0	0	0	0	0	0	0	0	0	0	0	0	0	0	0	0	0	0
8	1	1	0	0	0	0	0	0	0	0	1	1	1	1	0	0	0	1	1	1	1	1	1	1	0
9	1	1	0	0	0	0	0	0	0	0	1	1	1	1	1	1	1	1	1	1	1	1	1	1	0
10	1	1	0	0	0	0	0	0	0	0	1	1	1	1	1	1	1	1	1	1	1	1	1	1	0
11	1	1	0	0	0	0	0	0	0	0	1	1	1	1	0	0	0	1	1	1	1	1	1	1	0
12	1	1	0	0	0	0	0	0	0	0	0	0	0	0	0	0	0	0	0	0	0	0	0	0	0
13	1	1	0	0	0	0	0	0	0	0	0	0	0	0	0	0	0	1	1	1	1	0	0	0	0
14	1	1	0	0	0	0	0	0	0	0	0	0	0	0	0	0	0	0	0	0	0	0	0	0	0
15	0	0	0	0	0	0	0	0	0	0	0	0	0	0	0	0	0	1	1	1	1	1	1	0	0
16	0	0	0	0	0	0	0	0	0	0	0	0	0	0	0	0	0	1	1	1	1	1	1	0	0
17	0	0	0	0	0	0	0	0	0	0	0	0	0	0	0	0	0	0	0	0	0	0	0	0	0
18	0	0	0	0	0	0	0	0	0	0	0	0	0	0	0	0	0	0	0	0	0	0	0	0	0
19	1	1	0	0	0	0	0	0	0	1	1	1	1	1	1	1	1	1	1	1	1	1	1	1	1
20	1	1	0	0	0	0	0	0	0	1	1	1	1	1	1	1	1	1	1	1	1	1	1	1	1
21	1	1	0	0	0	0	0	0	0	1	1	1	1	1	1	1	1	1	1	1	1	1	1	1	1
22	0	0	0	0	0	0	0	0	0	0	0	0	0	0	0	0	0	0	0	0	0	0	0	0	0
23	0	0	0	0	0	0	0	0	0	0	0	0	0	0	0	0	0	0	0	0	0	0	0	0	0
24	0	0	0	0	0	0	0	0	0	0	0	0	0	0	0	0	0	0	0	0	0	0	0	0	0
25	0	0	0	0	0	0	0	0	0	0	0	0	0	0	0	0	0	0	0	0	0	0	0	0	0
26	0	0	0	0	0	0	0	0	0	0	0	0	0	0	0	0	0	0	0	0	0	0	0	0	0
27	0	0	0	0	0	0	0	0	0	0	0	0	0	0	0	0	0	0	0	0	0	0	0	0	0
28	1	1	1	1	1	1	1	1	1	1	1	1	1	1	1	1	1	1	1	1	1	1	1	1	1
29	1	1	1	1	1	1	1	1	1	1	1	1	1	1	1	1	1	1	1	1	1	1	1	1	1
30	1	1	1	1	1	1	1	1	1	1	1	1	1	1	1	1	1	1	1	1	1	1	1	1	1
31	1	1	1	1	1	1	1	1	1	1	1	1	1	1	1	1	1	1	1	1	1	1	1	1	1
32	1	1	1	1	1	1	1	1	1	1	1	1	1	1	1	1	1	1	1	1	1	1	1	1	1
33	1	1	1	1	1	1	1	1	1	1	1	1	1	1	1	1	1	1	1	1	1	1	1	1	1
34	1	1	1	1	1	1	1	1	1	1	1	1	1	1	1	1	1	1	1	1	1	1	1	1	1
35	1	1	1	1	1	1	1	1	1	1	1	1	1	1	1	1	1	1	1	1	1	1	1	1	1
36	1	1	1	1	1	1	1	1	1	1	1	1	1	1	1	1	1	1	1	1	1	1	1	1	1

4.5.2 Transmission Constraints. First we consider the steady state case. There will be 14 Benders cuts at hours 10-23 that will be added to the revised master problem. SCUC has an operating cost of $920,769 due to the commitment of more expensive units such as 7, 12, 16, 17, 18, 25 and 26. The execution time is 30 sec that is higher than that in the base case.

Next, we consider both steady state and contingency constraints in the transmission subproblems. The SCUC solution represents the loss of line 109-110 with a total operating cost of $925,324.4 and an execution time of 37 sec. There are 24 Benders cuts at hours 1-24 for contingencies in addition to 14 Benders cuts for the steady state case. Unit 4 is committed at peak hour 18 and units 17 and 18 are committed for additional hours to supply loads at buses 110 and 112 due to the line outage. Now, we consider additional transmission violations given in Table 4.2. The table shows the hours with flow violation and the corresponding number of Benders cuts in each possible line outages.

Table 4.2 Cost and Transmission Violations

Outaged lines	Cost ($)	Number of Benders Cuts	Hours with flow violation
109-110	925,324.4	24	1-24
82-96	928,147.6	24	1-24
69-77	937,227.9	17	5, 9-24
19-20	944,241.4	24	1-24
56-58	927,617.5	14	10-23

We consider the following security cases:

Case I (*nc* = *1*): line 109-110 possibly on outage

Case II (*nc* = *2*): line 109-110 or 82-96 possibly on outage

Case III (*nc* = *3*): line 109-110 or 82-96 or 69-77 possibly on outage

Case IV (*nc* = *4*): line 109-110 or 82-96 or 69-77 or 19-20 possibly on outage

Case V (*nc* = *5*): line 109-110 or 82-96 or 69-77 or 19-20 or 56-58 possibly on outage

Table 4.3 shows the results for these five cases. We run the security constrained economic dispatch at hour 18 to check the feasibility of SCUC in Case V. The results satisfy generation limits as well as line flow limits which support our algorithm.

Table 4.3 Cost and Cuts for Transmission Contingencies

Contingency	Cost ($)	Number of Benders Cuts	Hours with violation
I	925,324.4	38	1-24
II	930,484.6	62	1-24
III	940,267.8	79	1-24
IV	955,596.8	103	1-24
V	958,731.1	117	1-24

4.5.3 Voltage Constraints. First we solve the steady state transmission and voltage subproblems at each hour. There are 14 transmission Benders cuts at hours 10-23 that are added to the revised master problem. Also, there are 14 voltage Benders cuts at hours 10-23 that are added to the revised master problem. The total operating cost of $924,417.4 is increased now, due to the commitment of more expensive units 1, 6, 10 and 11. The execution time is 36 sec.

Next, we study both steady state and contingency transmission constraints in voltage and transmission subproblems considering the possible loss of line 109-110. The total operating cost is increased to $930,250.4 with expensive units 6, 10 and 17 for additional hours to accommodate voltage and transmission security constraints. The execution time is 49 sec and there are 14 voltage Benders cuts at hours 10-23 that are added to the revised master problem, in addition to 14 steady state voltage Benders cuts introduced above, and 38 transmission Bender cuts discussed earlier in Section 4.5.2. Table 4.4 shows the hours with voltage violations and the corresponding number of voltage Benders cuts in each possible line outage.

Table 4.4 Voltage Violations and Benders Cuts

Outaged line	Cost ($)	Number of Benders Cuts	Hours of Voltage Violation
109-110	930,250.4	14	10-23
82-96	932,695.3	14	10-23
69-77	942,593.3	17	1,9-24
19-20	964,491.4	17	1,9-24
56-58	932,799.1	14	10-23

The five contingency cases, discussed in Section 4.5.2, are adopted in this section to show the effect of voltage constraints on SCUC. In Case V with *nc=5*, we consider *5*24* contingency transmission subproblems and *5*24* contingency voltage subproblems in addition to *24* steady state transmission subproblems and *24* steady state voltage subproblems. For each contingency case, the revised production cost for SCUC is given in Table 4.5.

Table 4.5 Costs and Benders Cuts for Contingency Cases

Contingency Case	Total Cost ($)	Number of Transmission Benders Cuts	Number of Voltage Benders Cuts	Total Number of Benders Cuts
I	930,250.4	38	28	66
II	935,265.8	62	42	104
III	943,615.7	79	59	138
IV	964,806.3	103	76	179
V	964,806.3	117	90	207

Table 4.6 shows the scheduling results for Case V where five possible line outages are considered. The cost is $964,806.3 and the execution time is 400 sec as more expensive units 1, 4, 5, 6 and 11 (in comparison with Table 4.1) are committed for additional hours to satisfy voltage constraints.

Table 4.6 SCUC in Case V

No.	Hours (0-24)																							
1	0	1	0	0	0	0	0	0	1	1	1	1	1	1	1	1	1	1	1	1	1	1	1	1	
2	0	0	0	0	0	0	0	0	0	0	0	0	0	0	0	0	0	0	0	0	0	0	0	0	0
3	0	0	0	0	0	0	0	0	0	0	0	0	0	0	0	0	0	0	0	0	0	0	0	0	0
4	0	0	0	0	0	0	0	0	0	0	0	0	0	0	0	0	0	0	1	0	0	0	0	0	0
5	0	0	0	0	0	0	0	0	0	0	0	0	0	0	0	0	0	0	1	0	0	0	0	0	0
6	0	0	0	0	0	0	0	0	0	1	1	1	1	1	1	1	1	1	1	1	1	1	1	1	1
7	0	0	0	0	0	0	0	0	0	0	0	0	0	0	0	0	0	0	1	0	0	0	0	0	0
8	1	1	0	0	0	0	0	0	0	0	0	0	1	1	1	1	1	1	1	1	0	0	0	0	0
9	1	1	0	0	0	0	0	0	0	0	0	0	1	1	1	1	1	1	1	1	0	0	0	0	0
10	1	1	0	0	0	0	0	0	0	0	0	0	0	0	0	0	0	0	0	0	0	0	0	0	0
11	1	1	1	1	1	1	1	1	1	1	1	1	1	1	1	1	1	1	1	1	1	1	1	1	1
12	1	1	1	1	1	1	1	1	1	1	0	0	0	0	0	0	1	1	1	1	0	0	0	0	1
13	1	1	1	1	1	0	0	0	0	0	0	0	0	0	0	0	0	0	0	0	0	0	0	0	0
14	1	1	0	0	0	0	0	0	0	0	0	0	0	0	0	0	0	0	0	0	0	0	0	0	0
15	0	0	0	0	0	0	0	0	0	0	0	0	0	0	0	0	0	0	0	0	0	0	0	0	0
16	0	1	1	1	1	0	0	0	0	1	1	1	1	1	1	1	1	1	1	1	1	1	1	1	1
17	0	1	1	1	1	1	1	1	1	1	1	1	1	1	1	1	1	1	1	1	1	1	1	1	1
18	0	1	1	1	1	0	0	0	0	1	1	1	1	1	1	1	1	1	1	1	1	1	1	1	1
19	1	1	0	0	0	0	0	0	0	1	1	1	1	1	1	1	1	1	1	1	1	1	1	1	1
20	1	1	1	1	1	1	1	1	1	1	1	1	1	1	1	1	1	1	1	1	1	1	1	1	1
21	1	1	1	1	1	1	1	1	1	1	1	1	1	1	1	1	1	1	1	1	1	1	1	1	1
22	0	1	1	1	1	1	1	1	1	1	0	0	0	0	0	0	0	0	0	0	0	0	0	0	1
23	0	0	0	0	0	0	0	0	0	0	0	0	0	0	0	0	0	0	0	0	0	0	0	0	0
24	0	0	0	0	0	0	0	0	0	0	0	0	0	0	0	0	0	0	0	0	0	0	0	0	0
25	0	0	0	0	0	0	0	0	0	1	1	1	1	1	1	1	1	1	1	1	1	1	1	1	1
26	0	0	0	0	0	0	0	0	0	1	1	1	1	1	1	1	1	1	1	1	1	1	1	1	1
27	0	0	0	0	0	0	0	0	0	0	0	0	0	0	0	0	0	0	0	0	0	0	0	0	0
28	1	1	0	0	0	0	0	0	0	1	1	1	1	1	1	1	1	1	1	1	1	1	1	1	1
29	1	1	1	1	1	1	1	1	1	1	1	1	1	1	1	1	1	1	1	1	1	1	1	1	1
30	1	1	1	1	1	1	1	1	1	1	1	1	1	1	1	1	1	1	1	1	1	1	1	1	1
31	1	1	1	1	1	1	1	1	1	1	1	1	1	1	1	1	1	1	1	1	1	1	1	1	1
32	1	1	1	1	1	1	1	1	1	1	1	1	1	1	1	1	1	1	1	1	1	1	1	1	1
33	1	1	1	1	1	1	1	1	1	1	1	1	1	1	1	1	1	1	1	1	1	1	1	1	1
34	1	1	1	1	1	1	1	1	1	1	1	1	1	1	1	1	1	1	1	1	1	1	1	1	1
35	1	1	1	1	1	1	1	1	1	1	1	1	1	1	1	1	1	1	1	1	1	1	1	1	1
36	1	1	1	1	1	1	1	1	1	1	1	1	1	1	1	1	1	1	1	1	1	1	1	1	1

4.6 PRICE-BASED UNIT COMMITMENT

Competitive markets for energy and ancillary services are developing in different structures throughout the country. Only time and experience will show which market structures discourage gaming and yield efficient outcomes. The challenge is to determine which decisions can best be made by the ISO and which ones are to be made by market participants (e.g., GENCOs). Identifying fuel prices, demand, weather and hydro conditions can lead to accurate forecasts for power market prices. The accuracy in forecasting market prices is essential to price-based unit commitment. Moreover, it helps define GENCO's bidding strategies in order to maximize revenues. Quantities and prices for energy and reserves constitute the bidding strategy of a GENCO.

There are several models, such as Aurora, to forecast market prices in a competitive power marketplace. In this study, we use forecasted market prices to model price-based unit commitment for a GENCO. In this model, we examine different bidding strategies for energy, spinning and non-spinning reserve markets to determine the optimal bidding strategy that would maximize GENCOs' revenues.

4.7 PBUC PROBLEM DESCRIPTION

In a restructured power system, energy, spinning reserves and non-spinning reserves are offered as unbundled services. This section explains how GENCOs would commit their units in a restructured environment based on forecasted market prices and bidding strategies for each unit.

4.7.1 Interaction between GENCOS and ISO. Since the ISO and GENCO are two independent entities, we introduce a Benders decomposition which takes into account the ISO's control actions in generation scheduling. Here GENCOs solve price-based unit commitment without transmission security constraints and submit their bids to the ISO. Accordingly, the ISO will execute the congestion management and, if the solution is infeasible, will send a signal back to GENCOs with additional constrains as Benders cuts. GENCOs will then include these constraints into their price-based unit commitment (master problem) and solve it again. The interaction is shown in Figure 4.4.

4.8 GENCO'S BIDS

In a restructured system, GENCOs may bid for energy to supply load, spinning reserve, non-spinning reserve or submit a combination of bids. The bidding curves are explained as follows.

4.8.1 Energy Bids to Supply Loads. Each GENCO submits hourly price bids in a given time horizon to the ISO (or PX). In Figure 4.5, we assume that each generating unit in a GENCO may trade energy and submit incremental price bids for additional transactions in order to maximize its revenue.

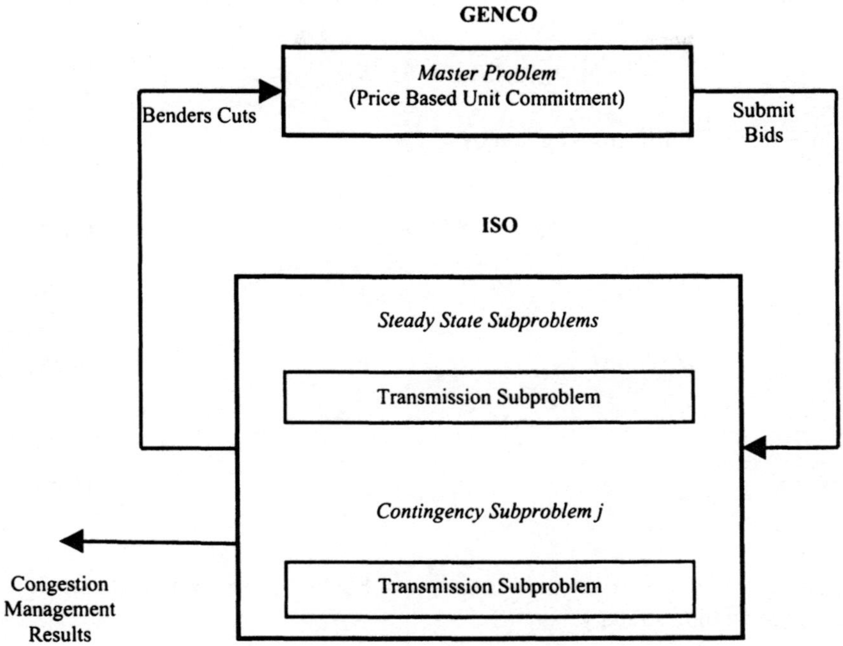

Figure 4.4 Interaction between GENCOs and ISO

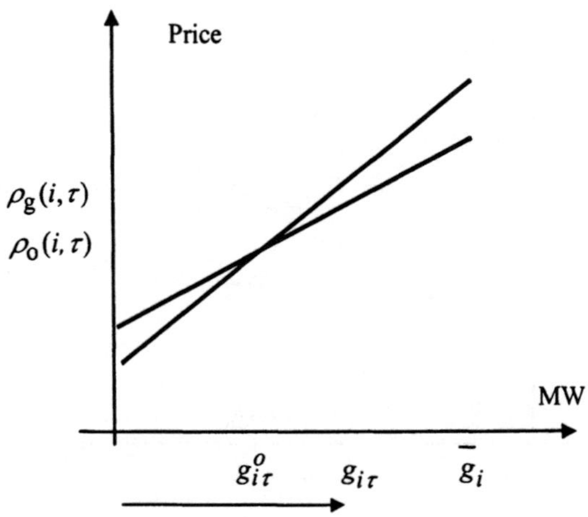

Figure 4.5 Bids for Energy to Supply Load

If the bid-in cost function of each generator at hour τ is given as:

$$F_i(g_{i\tau}) = a_{1i} + a_{2i} * g_{i\tau} + a_{3i} * g_{i\tau}^2 \tag{4.57}$$

Then, the incremental bid-in cost is given by:

$$\frac{\Delta F_i(g_{i\tau})}{\Delta g_{i\tau}} = a_{2i} + a_{3i} * g_{i\tau} + a_{3i} * g_{i\tau}^o \tag{4.58}$$

where

$$\Delta F_i(g_{i\tau}) = F_i(g_{i\tau}) - F_i(g_{i\tau}^o) \tag{4.59}$$

and

$$\Delta g_{i\tau} = g_{i\tau} - g_{i\tau}^o \tag{4.60}$$

Correspondingly, the bid-in price curve is shown in Figure 4.5 where the adjustable slope defines the bidding strategy.

4.8.2 Spinning and Non-Spinning Reserve Bids.

The ISO is the buyer of reserve capacity with an obligation to meet the reliability of the system. Based on the quantity and the location of system requirements, the ISO will select the most appropriate generating units with lowest bids. Spinning and non-spinning reserve bids may be illustrated as in Figures 4.6& 4.7, respectively.

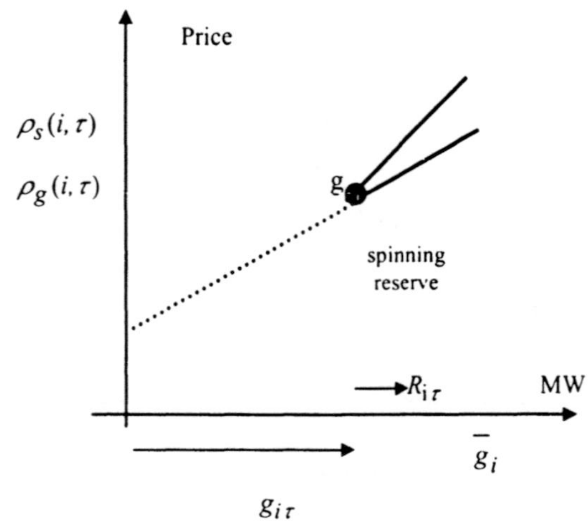

Figure 4.6 Spinning Reserve Bid

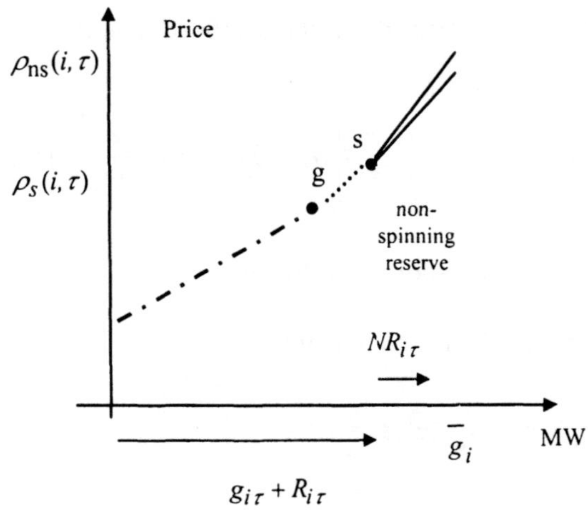

Figure 4.7 Non-Spinning Reserve Bid

In Figure 4.6, point g distinguishes between the energy supply and the spinning reserve supply. Here, the presumption is that the slope for spinning reserve bid is adjustable and GENCOs will try to optimize their revenues by adjusting the slope. Furthermore, point s in Figure 4.7 distinguishes between spinning and non-spinning reserves. Likewise, the slope for the spinning reserve bid is fixed now and the GENCO adjusts the bid for non-spinning reserve to optimize the revenues.

4.9 GENCO'S OBJECTIVE

Based on the cost and revenue discussion in Section 4.1 and for a given supply of energy, spinning and non-spinning reserves, and forecasted market price, each GENCO maximizes its profit at a certain hour τ as follows:

$$Maximize \sum_{\tau=1}\sum_{i=1}\left\{\begin{array}{c}\rho_g(i,\tau)*[g_{i\tau}I_{i\tau}+\chi_{i\tau}-g^o_{i\tau}]+\rho_s(i,\tau)*R_{i\tau}I_{i\tau}\\ -S_{i\tau}+\rho_{ns}(i,\tau)*NR_{i\tau}-\rho_\chi(i,\tau)\chi_{i\tau}\\ -F_i([g_{i\tau}+R_{i\tau}I_{i\tau}]+NR_{i\tau})+F_i(g^o_{i\tau})\end{array}\right\} \tag{4.61}$$

which is equivalent to,

$$Minimize \sum_{\tau}\sum_{i}\left\{\begin{array}{c}-\rho_g(i,\tau)*[g_{i\tau}I_{i\tau}+\chi_{i\tau}-g_{i\tau}]-\rho_s(i,\tau)*R_{i\tau}I_{i\tau}\\ +S_{i\tau}-\rho_{ns}(i,\tau)*NR_{i\tau}+\rho_\chi(i,\tau)\chi_{i\tau}\\ +F_i([g_{i\tau}+R_{i\tau}]I_{i\tau}+NR_{i\tau})-F_i(g^o_{i\tau})\end{array}\right\} \tag{4.62}$$

The start up cost of a unit depends on the length of time that the unit has been off and is defined as:

$$S_{i\tau} = I_{i\tau}[1 - I_{i(\tau-1)}]\left[\alpha_i + \beta_i\left(1 - exp\frac{-Z_{i\tau}^{off}}{\Theta_i}\right)\right] \tag{4.63}$$

subject to the following operational constraints:

a) Limits on bids offered by generating units

$$\underline{\rho}_g(i,\tau) \le \rho_g(i,\tau) \le \overline{\rho}_g(i,\tau) \tag{4.64}$$

$$\underline{\rho}_s(i,\tau) \le \rho_s(i,\tau) \le \overline{\rho}_s(i,\tau) \tag{4.65}$$

$$\underline{\rho}_{ns}(i,\tau) \le \rho_{ns}(i,\tau) \le \overline{\rho}_{ns}(i,\tau) \tag{4.66}$$

$$\rho_\chi(i,\tau) \le \rho_\chi(i,\tau) \le \overline{\rho}_\chi(i,\tau) \tag{4.67}$$

where $\underline{\rho}_g(i,\tau)$, $\overline{\rho}_\chi(i,\tau)$ are determined by marginal prices of generating units. $\underline{\rho}_\chi(i,\tau)$ may be set to zero and other limits will be determined from the forecasted market price.

b) GENCO's real power and reserve limits

$$\underline{g}_i \le \sum_i g_{i\tau} I_{i\tau} \le \overline{g}_i \tag{4.68}$$

$$R_o(\tau) \le \sum_i R_{i\tau} I_{i\tau} \le \overline{R}(\tau) \tag{4.69}$$

$$NR_o(\tau) \le \sum_i NR_{i\tau} \le \overline{NR}(\tau) \tag{4.70}$$

$$\sum_i \chi_{i\tau} \le \sum_i g_{i\tau}^o \tag{4.71}$$

$$I_{i\tau} g_{i\tau} + \chi_{i\tau} \ge g_{i\tau}^o \tag{4.72}$$

$$\underline{g}_i \le \{g_{i\tau} + R_{i\tau}\} I_{i\tau} + NR_{i\tau} \le \overline{g}_i \tag{4.73}$$

$$0 \leq \chi_{i\tau} \leq g_{i\tau}^{o} \tag{4.74}$$

$$R_{i\tau} I_{i\tau} \leq r_{si}(\tau) I_{i\tau} \tag{4.75}$$

where , $r_{si}(\tau) = min\{10 * MSR_i, \bar{g}_i - g_{i\tau}\}$.

$$NR_{i\tau} \leq r_{oi}(\tau) \tag{4.76}$$

where $r_{oi}(\tau) = \begin{cases} q_i, & \text{if unit } i \text{ is OFF} \\ r'_{si}(\tau), & \text{if unit } i \text{ is ON} \end{cases}$

and $r'_{si}(\tau) = min\{10 * MSR_i, \bar{g}_i - g_{i\tau} - R(i,\tau)\}$

c) Thermal unit minimum start up/down times

$$(Z_{i(\tau-1)}^{on} - T_i^{on}) * (I_{i(\tau-1)} - I_{i\tau}) \geq 0 \tag{4.77}$$

$$(Z_{i(\tau-1)}^{off} - T_i^{off}) * (I_{i(\tau-1)} - I_{i\tau}) \geq 0 \tag{4.78}$$

d) Ramp rate limits

$$g_{i\tau} - g_{i(\tau-1)} \leq UR_i \quad \text{as unit } i \text{ ramps up} \tag{4.79}$$

$$g_{i(\tau-1)} - g_{i\tau} \leq DR_i \quad \text{as unit } i \text{ ramps down} \tag{4.80}$$

e) Fuel constraints

$$FL \leq \sum_{\tau} \sum_{i \in A} f_{ci} * [F_i([g_{i\tau} + R_{i\tau}] I_{i\tau} + NR_{i\tau}) + S_{i\tau}] \leq FU \tag{4.81}$$

where $i \in A$ corresponds to all fuel-constrained units in group A.

f) System emission limit

$$\sum_{\tau} \sum_{i} H_i([g_{i\tau} + R_{i\tau}] I_{i\tau} + NR_{i\tau}) \leq EMS_t \tag{4.82}$$

where several emission types (e.g. $SO2$, NO_X) may be considered.

g) Area emission limit

$$\sum_{\tau}\sum_{i\in B} H_i([\,g_{i\tau}+R_{i\tau}\,]\,I_{i\tau}+NR_{i\tau}) \le EMA_t \tag{4.83}$$

where $i \in B$ corresponds to all units in the constrained emission area B and several emission types can be considered.

4.10 PBUC SOLUTION METHODOLOGY

One of the potential approaches for solving the mixed-integer programming problem is the Lagrangian relaxation. Again, the relaxed problem is decomposed into N subproblems representing each unit. Dynamic programming is used to search the optimal commitment for single units. Lagrangian multipliers are updated based on violations of system constraints. The convergence criterion is:

$$\frac{2*(\,Profit^{\,j+1} - Profit^{\,j}\,)}{(\,Profit^{\,j+1} + Profit^{\,j}\,)} \le \Delta \tag{4.84}$$

where $Profit^{\,j+1}$ is the profit at iteration $j+1$ and $Profit^{\,j}$ is the profit at iteration j. The convergence criterion is satisfied if Δ is within the given limits. There is a trade-off between choosing a smaller Δ and the number of iterations. The smaller the Δ, the larger the number of iterations will be. For some cases, the problem may not converge if Δ is too small.

The Lagrangian function for PBUC is given next where terms that are independent of the decision variable are dropped:

$$\Gamma\,\{g_{i\tau}, R_{i\tau}, NR_{i\tau}, \chi_{i\tau}, I_{i\tau}, \rho_g(i,\tau), \rho_s(i,\tau), \rho_{ns}(i,\tau), \rho_\chi(i,\tau), \lambda_1(\tau), \lambda_2(\tau), \lambda_3(\tau), \lambda_4(\tau), \lambda_f, \lambda_s, \lambda_a\,\}$$

$$= \sum_{\tau}\sum_{i}\left\{\begin{aligned}&-\rho_g(i,\tau)*[g_{i\tau}\,I_{i\tau} + \chi_{i\tau} - g^{o}_{i\tau}] - \rho_s(i,\tau)*R_{i\tau}\,I_{i\tau} + S_{i\tau}\\ &-\rho_{ns}(i,\tau)*NR_{i\tau} + \rho_\chi(i,\tau)\,\chi_{i\tau} + F_i([g_{i\tau} + R_{i\tau}]\,I_{i\tau} + NR_{i\tau})\end{aligned}\right\}$$

$$+\sum_{\tau}\lambda_1(\tau)\sum_{i} g_{i\tau} I_{i\tau} + \sum_{\tau}\lambda_2(\tau)\sum_{i} R_{i\tau} I_{i\tau}$$

$$+\sum_{\tau}\lambda_3(\tau)\sum_{i} NR_{i\tau} + \sum_{\tau}\lambda_4(\tau)\sum_{i} \chi_{i\tau}$$

$$+\lambda_f \sum_{\tau} \sum_{i \in A} f_{ci} * [F_i([g_{i\tau} + R_{i\tau}] I_{i\tau} + NR_{i\tau}) + S_{i\tau}] \tag{4.85}$$

$$+\lambda_s \sum_{\tau} \sum_{i} H_i([g_{i\tau} + R_{i\tau}] I_{i\tau} + NR_{i\tau})$$

$$+\lambda_a \sum_{\tau} \sum_{i \in B} H_i([g_{i\tau} + R_{i\tau}] I_{i\tau} + NR_{i\tau})$$

subject to bus constraints (4.12-4.16). The optimization problem in $k+1^{th}$ iteration can be decoupled into N subproblems, each corresponding to the optimal unit commitment of individual units over the entire study period.

At this stage, $g_{i\tau}, R_{i\tau}, NR_{i\tau}$ and $\chi_{i\tau}$ are set equal to power generation corresponding to Lagrangian multipliers $\lambda_1(\tau), \lambda_2(\tau), \lambda_3(\tau),$ *and* $\lambda_4(\tau)$. Consequently, integer variables $I_{i\tau}$, $\tau = 1,....,n_\tau$ are the only unknown factors. We adopt a dynamic programming model to find the suitable values of $I_{i\tau}$, which would minimize (4.85).

Once generating units are committed, we solve the economic dispatch similar to that of Section 4.4.1. Based on the Lagrangian function defined in (4.85), the corresponding optimal values of $g_{i\tau}, R_{i\tau}, NR_{i\tau}$ and $\chi_{i\tau}$ for each unit are derived as follows:

$$\begin{aligned}
\frac{\partial \Gamma}{\partial g_{i\tau}} &= -\rho_g(i,\tau) + \frac{\partial F_i}{\partial g_{i\tau}} + \lambda_1(\tau) = 0.0 \\
\frac{\partial \Gamma}{\partial R_{i\tau}} &= -\rho_s(i,\tau) + \frac{\partial F_i}{\partial R_{i\tau}} + \lambda_2(\tau) = 0.0 \\
\frac{\partial \Gamma}{\partial NR_{i\tau}} &= -\rho_{ns}(i,\tau) + \frac{\partial F_i}{\partial NR_{i\tau}} + \lambda_3(\tau) = 0.0 \\
\frac{\partial \Gamma}{\partial \chi_{i\tau}} &= -\rho_g(i,\tau) + \chi_{i\tau} + \lambda_4(\tau) = 0.0
\end{aligned} \tag{4.86}$$

Based on (4.85), prices are written as follows:

$$
\begin{aligned}
\rho_g(i,\tau) &= a_{2i} + a_{3i}*g^o_{i\tau} + [1+\lambda_1(\tau)]*a_{3i}*[g_{i\tau} + \chi_{i\tau}] \\
\rho_s(i,\tau) &= \rho_g(i,\tau) + [1+\lambda_2(\tau)]*a_{3i}*R_{i\tau} \\
\rho_{ns}(i,\tau) &= \rho_s(i,\tau) + [1+\lambda_3(\tau)]*a_{3i}*NR_{i\tau} \\
\rho_\chi(i,\tau) &= \rho_o(i,\tau) - [1+\lambda_4(\tau)]*a_{3i}*\chi_{i\tau}
\end{aligned}
\tag{4.87}
$$

where,

$$\rho_o(i,\tau) = a_{2i} + 2*a_{3i}*g^o_{i\tau}$$

From (4.86) and (4.87), we get the following linear equations:

$$
\begin{aligned}
&[1-\lambda_1(\tau)]*g_{i\tau} + 2*R_{i\tau} + 2*NR_{i\tau} - [1+\lambda_1(\tau)]*\chi_{i\tau} = g^o_{i\tau} - \lambda_1(\tau)/a_{3i} \\
&[1-\lambda_1(\tau)]*g_{i\tau} + [1-\lambda_2(\tau)]*R_{i\tau} + 2*NR_{i\tau} - [1+\lambda_1(\tau)]*\chi_{i\tau} = g^o_{i\tau} - \lambda_2(\tau)/a_{3i} \\
&[1-\lambda_1(\tau)]*g_{i\tau} + [1-\lambda_2(\tau)]*R_{i\tau} + [1-\lambda_3(\tau)]*NR_{i\tau} - [1+\lambda_1(\tau)]*\chi_{i\tau} = g^o_{i\tau} - \lambda_3(\tau)/a_{3i} \\
&[1+\lambda_1(\tau)]*g_{i\tau} + [2+\lambda_1(\tau)+\lambda_4(\tau)]*\chi_{i\tau} = g^o_{i\tau} + \lambda_4(\tau)/a_{3i}
\end{aligned}
\tag{4.88}
$$

Then, (4.88) may be re-written in a matrix form as follows:

$$
\begin{bmatrix}
1-\lambda_1(\tau) & 2 & 2 & -1-\lambda_1(\tau) \\
1-\lambda_1(\tau) & 1-\lambda_2(\tau) & 2 & -1-\lambda_1(\tau) \\
1-\lambda_1(\tau) & 1-\lambda_2(\tau) & 1-\lambda_3(\tau) & -1-\lambda_1(\tau) \\
1+\lambda_1(\tau) & 0 & 0 & 2+\lambda_1(\tau)+\lambda_4(\tau)
\end{bmatrix}
\begin{bmatrix}
g_{i\tau} \\ R_{i\tau} \\ NR_{i\tau} \\ \chi_{i\tau}
\end{bmatrix}
=
\begin{bmatrix}
g^o_{i\tau} - \dfrac{\lambda_1(\tau)}{a_{3i}} \\
g^o_{i\tau} - \dfrac{\lambda_2(\tau)}{a_{3i}} \\
g^o_{i\tau} - \dfrac{\lambda_3(\tau)}{a_{3i}} \\
g^o_{i\tau} + \dfrac{\lambda_4(\tau)}{a_{3i}}
\end{bmatrix}
\tag{4.89}
$$

The above set of linear equations is solved by substitution, and optimal values of, $g_{i\tau}$, $R_{i\tau}$, $NR_{i\tau}$ and $\chi_{i\tau}$ are determined as a function of Lagrangian multipliers $\lambda_1(\tau)$, $\lambda_2(\tau)$, $\lambda_3(\tau)$, *and* $\lambda_4(\tau)$ as follows:

$$g_{i\tau} = [(1+\lambda_1(\tau)) / (1+\lambda_2(\tau)] * \Big\{ \{-2*\lambda_1(\tau)*g^o_{i\tau} + 2*(1+\lambda_1(\tau))*[\lambda_2(\tau)-\lambda_1(\tau)] / [(1+\lambda_2(\tau))*a_{3i}] + 2*(1+\lambda_1(\tau))*[\lambda_2(\tau)-\lambda_3(\tau)] / [(1+\lambda_3(\tau))*a_{3i}] + \lambda_1(\tau)*(1+\lambda_1(\tau))/a_{3i} + \lambda_4(\tau)*(1+\lambda_4(\tau))a_{3i}\} / \{3+\lambda_1(\tau)+\lambda_4(\tau) - \lambda_1(\tau)\lambda_4(\tau)\} \Big\} - [2/(1-\lambda_1(\tau))]*[\lambda_2(\tau)-\lambda_1(\tau)] / [(1+\lambda_2(\tau))*a_{3i}] - [2/(1-\lambda_1(\tau))]*[\lambda_2(\tau)-\lambda_3(\tau)] / [(1+\lambda_3(\tau))*a_{3i}] + [g^o_{i\tau} - \lambda_1(\tau)/a_{3i}]/[1-\lambda_1(\tau)]$$

$$R_{i\tau} = [\lambda_2(\tau)-\lambda_1(\tau)] / [(1+\lambda_2(\tau))*a_{3i}] \tag{4.90}$$

$$NR_{i\tau} = [\lambda_2(\tau)-\lambda_3(\tau)] / [(1+\lambda_3(\tau))*a_{3i}]$$

$$\chi_{i\tau} = \{-2*\lambda_1(\tau)*g^o_{i\tau} + 2*(1+\lambda_1(\tau))*[\lambda_2(\tau)-\lambda_1(\tau)] / [(1+\lambda_2(\tau))*a_{3i}] + 2*(1+\lambda_1(\tau))*[\lambda_2(\tau)-\lambda_3(\tau)]/[(1+\lambda_3(\tau))*a_{3i}] + \lambda_1(\tau)*(1+\lambda_1(\tau))/a_{3i} + \lambda_4(\tau)*(1+\lambda_4(\tau))/a_{3i}\} / \{3+\lambda_1(\tau)+\lambda_4(\tau)-\lambda_1(\tau)\lambda_4(\tau)\}$$

In a GENCO, we use the following equations to determine the amount of power to be sold to supply loads, total power to be purchased to supply bilateral contracts, as well as spinning and non-spinning reserves to be sold:

$$\text{Total power to be sold} = \sum_{\tau} \sum_{i} \{ g_{i\tau} + \chi_{i\tau} - g^o_{i\tau} \}$$

$$\text{Total power to be purchased} = \sum_{\tau} \sum_{i} \{ \chi_{i\tau} \}$$

$$\text{Total spinning reserve to be sold} = \sum_{\tau} \sum_{i} \{ R_{i\tau} \} \tag{4.91}$$

$$\text{Total non-spinning reserve to be sold} = \sum_{\tau} \sum_{i} \{ NR_{i\tau} \}$$

The incremental/decremental adjustment bids are used by the ISO for congestion management. The slope for incremental/decremental adjustment bids offered by unit i in a GENCO is calculated as follows:

$$m_g(i,\tau) = [1+\lambda_1(\tau)]*a_{3i} \tag{4.92}$$

CHAPTER V

COORDINATION BETWEEN LONG-TERM AND SHORT-TERM GENERATION SCHEDULING

5.1 INTRODUCTION

This chapter extends short-term generation scheduling constraints to the solution of long-term generation maintenance scheduling. The results indicate that it is essential to check the short-term operational constraints prior to implementing long-term maintenance scheduling.

The hierarchical structure for generation maintenance scheduling is shown in Figure 5.1. In this figure, long-term scheduling (LTS) applies to the long-range maintenance planning and fuel budgeting which ought to be coordinated with short-term scheduling. Short-term scheduling (STS) will minimize the operation cost in hourly, daily or weekly horizons for a given value of the load. STS will meet fuel supply and consumption limits, unit I/O characteristics, GENCO's requirements and hourly demands and contracts.

LTS in a GENCO is not an independent problem. In LTS, maintenance scheduling is used to determine specific windows for generation maintenance scheduling. Using the LTS window, STS will examine the possible commitment of available generating units. The solution is submitted to the ISO to check system constraints (Figure 5.1). The proper coordination between LTS and STS in a GENCO is imperative to ensure that the LTS output can be used to facilitate STS. Other analytical methods were suggested in the literature for the coordination between LTS and STS which focus mostly on the fuel resource scheduling problem [Gar94, Lee92, Aga95, Wet89].

This chapter will link subjects discussed in Chapters III and IV. The chapter consists of four major sections. Section 5.2 and 5.3 outline the LTS and STS formulations consecutively. The proposed correlated decision function for LTS and STS is described in Sections 5.4 and 5.5. Section 5.6 illustrates the results of applying the proposed method to IEEE-RTS [see Appendix A].

5.2 LONG-TERM GENERATION MAINTENANCE FORMULATION

As discussed in Chapter III, accounting for the interaction between fuel restrictions, allowable emission and ISO constraints will greatly increase the complexity of the LTS problem, resulting in a large optimization problem. The LTS problem has discrete decision variables related to maintenance scheduling, and continuos variables representing fuel allocations and utilization of generating units in a GENCO. The scheduling may involve a time horizon of one or two years that can be divided into window intervals (weeks).

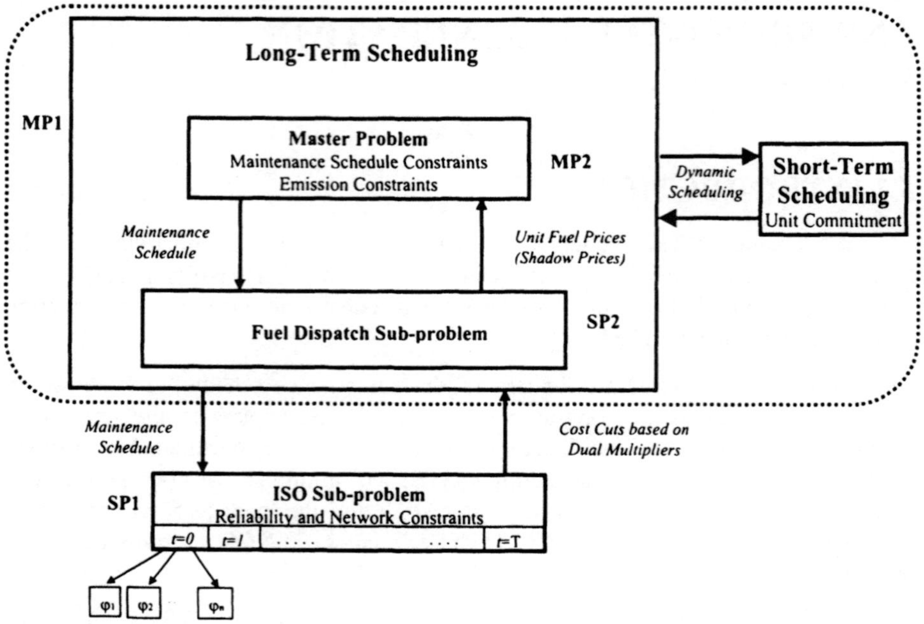

Figure 5.1 Dynamic Scheduling of the Generation Maintenance

LTS in given intervals will satisfy weekly fuel supply limits, unit maintenance constraints, and unit characteristics. The formulation of LTS was discussed thoroughly in Chapter III and is listed as follows:

$$Min \sum_{t}\left\{\sum_{i} C_{it}(1-x_{it})\right\}+\sum_{i}\sum_{m}\sum_{t} c_{mt}\phi_{imt}+\sum_{t}\chi_t E\left\{\sum_{i} r_{it}\right\}$$

St.

$$x_{it}=1 \qquad for\ t\le e_i\ or\ t\ge l_i+d_i$$

$$x_{it}=0 \qquad for\ s_i\le t\le s_i+d_i$$

$$x_{it}\in\{0,1\} \qquad for\ e_i\le t\le l_i$$

1. crew availability   *3. seasonal limitations*

2. resources availability   *4. desirable schedule*

$$\sum_t \sum_i H_i(E\{g_{it}\}) x_{it} \le EMS$$

$$\sum_t \sum_{i \in A} H_i(E\{g_{it}\}) x_{it} \le EMA$$

$$\sum_m \phi_{imt} \le K_i x_{it} \qquad \text{for } \forall i \text{ and } \forall t$$

$$\sum_i \phi_{imt} \le W_{mt} \qquad \text{for } \forall m \text{ and } \forall t$$

$$\sum_i \sum_{t \in month\ j} \phi_{imt} \le W_{mj} \qquad \text{for } \forall m \text{ and } \forall month\ j \tag{5.1}$$

$$\sum_i \sum_{t \in year\ v} \phi_{imt} \le W_{mv} \qquad \text{for } \forall m \text{ and } \forall year\ v$$

$$E\{g_{it}\} \le \sum_m \frac{\phi_{imt}}{h_{im}} \qquad \text{for } \forall i \text{ and } \forall t$$

The reader is required to review the content of Chapter III for the solution of (5.1).

5.3 SHORT-TERM FORMULATION

The difference between LTS and STS in a GENCO is in the designated time horizon and time increment for the solution of the maintenance problem. In LTS, the horizon of study is one or two years with weekly increments (Δt), but in STS the horizon of study is weeks with hourly increments ($\Delta\tau$). Here, the appropriate window for STS is determined by a LTS. Then, outputs of LTS, variables x and ϕ are passed on to the STS formulation. These variables are represented by constraints (5.11) and (5.14) below.

Unit commitment was covered in detail in Chapter IV. We discussed both cost-based and the price-based solutions. Since in the case of LTS we predict a given value of the load for each window, we also use the cost-based unit commitment formulation here in which the value of the load is assumed known. The STS formulation which encompasses unit commitment constraints in addition to LTS variables x and ϕ is given below:

$$PC = \sum_i \sum_\tau \left[I_{i\tau} F_i(g_{i\tau}) + S_{i\tau} \right] \tag{5.2}$$

The production cost $F_i(g_{i\tau})$ is calculated as the product of the heat rate (MBTU/Hr) and the unit's fuel cost ($/MBTU). The start up cost of a unit depends on how long the unit has been off, and is defined as

$$S_{i\tau} = I_{i\tau}[1 - I_{i(\tau-1)}]\left[\alpha_i + \beta_i\left(1 - \exp\frac{-Z_{i\tau}^{off}}{\theta_i}\right)\right]$$

The constraints of the optimization problem are:

a) System real power balance

$$\sum_i I_{i\tau} g_{i\tau} = P_D(\tau) \qquad \tau = 1,\dots,\ t \tag{5.3}$$

b) System spinning reserve requirements

$$\sum_i r_{si}(\tau) I_{i\tau} \geq R_s(\tau) \qquad \tau = 1,\dots,\ t \tag{5.4}$$

where, $r_{si}(\tau) = min\{10 * MSR_i, \overline{g}_i - g_{i\tau}\}$. The spinning reserve requirement, $R_S(\tau)$, is typically defined as a base component plus a fraction of the load and a fraction of the high operating limit of the largest on-line unit.

c) System operating reserve requirements

$$\sum_i r_{oi}(\tau) I_{i\tau} \geq R_o(\tau) \qquad \tau = 1,\dots,\ t \tag{5.5}$$

where, $r_{oi}(\tau) = \begin{cases} q_i, & \text{if unit } i \text{ is OFF} \\ r_{si}(\tau), & \text{if unit } i \text{ is ON} \end{cases}$

Interruptible loads are usually added to the operating reserve capacity of units (left hand side of (5.4)). The operating reserve requirement, $R_0(\tau)$, is commonly defined similar to $R_s(\tau)$ as a function of a base component plus a fraction of the load and a fraction of the high operating limit of the largest on-line unit.

d) Unit generation limits

$$\underline{g}_i \leq g_{i\tau} \leq \overline{g}_i \qquad \tau = 1,\dots,\ t \qquad \forall i \tag{5.6}$$

e) Thermal unit minimum start up/down times

$$(Z_{i(\tau-1)}^{on} - T_i^{on}) * (I_{i(\tau-1)} - I_{i\tau}) \geq 0 \tag{5.7}$$

$$(Z_{i(\tau-1)}^{off} - T_i^{off}) * (I_{i(\tau-1)} - I_{i\tau}) \geq 0 \tag{5.8}$$

f) Ramp rate limits

$$g_{i\tau} - g_{i(\tau-1)} \leq UR_i \qquad \textit{as unit i ramps up} \tag{5.9}$$

$$g_{i(\tau-1)} - g_{i\tau} \leq DR_i \qquad \textit{as unit i ramps down} \tag{5.10}$$

g) Fuel constraints

$$FL \leq \sum_{\tau} \sum_{i \in A} f_{ci} * [I_{i\tau} F_i(g_{i\tau}) + S_{i\tau}] \leq \sum_{i \in A} \phi_{imt} \tag{5.11}$$

where $i \in A$ corresponds to all fuel-constrained units in group A. The first item $f_{ci} I_{i\tau} F_i(g_{i\tau})$ is operating fuel and $f_{ci} S_{i\tau}$ is additional fuel for start up.

h) System emission limit

$$\sum_{\tau} \sum_{i} H_i(g_{i\tau}) I_{i\tau} \leq EMS_t \tag{5.12}$$

where several emission types (e.g. $SO2$, NO_X) are considered.

i) Area emission limit

$$\sum_{\tau} \sum_{i \in B} H_i(g_{i\tau}) I_{i\tau} \leq EMA_t \tag{5.13}$$

where $i \in B$ corresponds to all units in the constrained emission area B where several emission types can be considered.

j) Maintenance constraints

$$I_{i\tau} \leq x_{it} \qquad \textit{for } \tau \in t \tag{5.14}$$

We may apply Lagrangian relaxation (LR), discussed in Chapter IV, to solve the STS problem. Constraint (5.14) links the maintenance variable x_{it} with the commitment state. In order to set proper emission caps (EMA_t and EMS_t) in STS, emission caps in LTS are distributed in proportion to the weekly peak load. The fuel scheduling problem was solved earlier using a primal approach which employed weekly targets for coordination. LTS allocates a sequence of weekly fuel targets as fuel constraints in STS given by (5.11). In the case of multiple constrained fuels, a dual coordination is preferable for fuel scheduling.

5.4 DYNAMIC SCHEDULING OF LTS

Facing future uncertainties, the most effective maintenance strategy is based on dynamic scheduling. This strategy introduces a decision, conditional on past realizations, to calculate future maintenance schedules such that the expected cost is minimized. It is clear that the foremost advantage of this strategy lies on the flexibility to adjust the maintenance decision adaptively when additional information is made available. To account for this, LTS and STS coordination is shown in Figure 5.2.

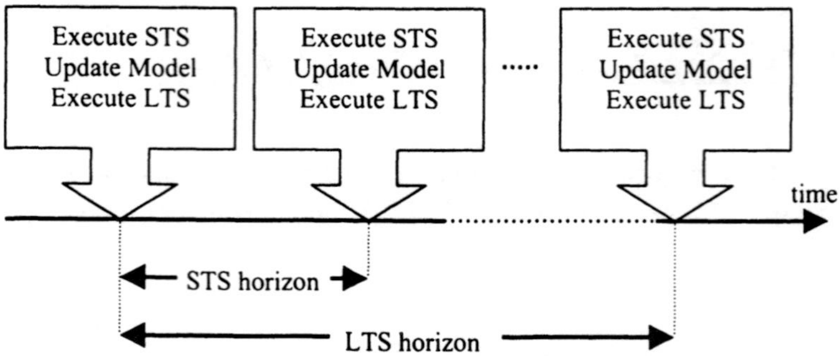

Figure 5.2 Proposed Dynamic Scheduling

In Figure 5.2, each STS execution is preceded by LTS while in practice this may be deemed unnecessary at certain times. Whether this will be the case will depend on the system performance in the course of the prior week and new pieces of information that may become available based on the STS solution.

LTS is dependent on many forecast parameters such as load demand and generating units' availability. One can only speculate on future values of these parameters based on today's facts; hence, forced outage rates of generation units should be updated according to the end of the week information. In this chapter, we use a two-state continuos-time Markov model [Oli89] for each generating unit and transmission line. For example, the associated conditional probabilities for a generating unit are:

$$p(\varphi_t = up \mid \varphi_{t=to} = up) = p + qe^{-(\mu+\lambda)(t-to)}$$

$$p(\varphi_t = down \mid \varphi_{t=to} = up) = q - qe^{-(\mu+\lambda)(t-to)}$$

$$p(\varphi_t = up \mid \varphi_{t=to} = down) = p - pe^{-(\mu+\lambda)(t-to)}$$

$$p(\varphi_t = down \mid \varphi_{t=to} = down) = q + pe^{-(\mu+\lambda)(t-to)} \quad (5.15)$$

where, p, q, μ, λ represent the generating unit's steady state availability, steady state unavailability, repair rate in $(\text{day})^{-1}$, and failure rate in $(\text{day})^{-1}$ respectively. Accordingly, forced outage rates of all components in the system will be calculated using (5.15) at weekly peak loads. These updated values of forced outages are assigned to constraints in the LTS calculation.

The proper scheduling of fuel is important in that it is desirable to burn an appropriate amount of fuel required by contracts. Therefore, it is important to periodically update the fuel usage schedule to allow for changes in demand. The weekly amount of constrained fuel burn is determined by prorating the amount of remaining constrained annual fuel by the ratio of the predicted weekly generation to the predicted annual generation.

The proposed dynamic scheduling algorithm for the coordination between LTS and STS is depicted in Figure 5.3. Based on the STS solution, LTS is updated every week as necessary. The deviation of STS variables from their anticipated values is used as the criterion. In this chapter, we use the reserve capacity as an indicator. If GENCO's reserve at any week is below requirements, the LTS solution will be updated.

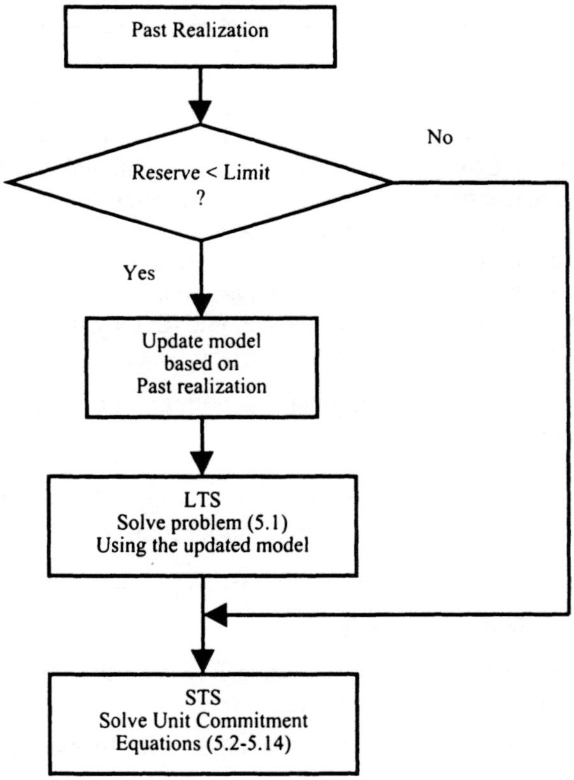

Figure 5.3 Dynamic Scheduling Algorithm

5.5 ISO CONSTRAINTS

The formulation of TRANSCO's constraints, monitored by the ISO, is the same as that in preceding chapters:

$$
\begin{aligned}
&Sf + g + r = d(\varphi) && \forall t \\
&g \le \bar{g}(\varphi).x && \forall t \\
&r \le d(\varphi) && \forall t \\
&|f| \le \bar{f}(\varphi) && \forall t
\end{aligned}
\qquad (5.16)
$$

5.6 CASE STUDIES

We apply the same system as the one in Chapter IV. For the purpose of STS, the typical hourly load data for a day are listed in Table A3 of Appendix A. Generating unit data for unit commitment such as the production cost function and the minimum up/down time are listed in Tables A6 and A7 of Appendix A.

5.6.1 Monte-Carlo Simulation. In order to test the proposed dynamic scheduling, various randomly generated scenarios, reflecting future uncertainties, are considered.

Since the availability of each generating unit due to forced outages is uncertain, we describe conditional probabilities associated with this uncertainty by (5.15). To simulate the GENCO's state realization, a conditional Monte-Carlo sampling (i.e. sampling sequentially conditional on realization of past samples) is used to randomly generate future scenarios. Details of the Monte-Carlo sampling algorithm is found in Appendix D. Sequential sampling will allow the dynamic scheduling strategy to take its course. In this application, 1000 randomly generated scenarios (i.e. each scenario specifies the realized GENCO's states over a 12-week study) are used to evaluate the dynamic maintenance schedule.

The Monte-Carlo sampling algorithm with generalized regression is implemented as follows:

1. Calculate the expected production cost using LTS over the specified simulation horizon. Let $E\{Z\}$ be the expected production cost.
2. Sample a generation capacity vector $\{\bar{g}_j^i\}$ where $\bar{g}_j^i$ is the sampled generation capacity in the i-th sample.
3. Calculate the production cost using LTS over the specified simulation horizon. Since the set of available units is known from Step 2, the calculation is essentially a deterministic simulation. Let Z^i be the production cost of the i-th sample.
4. The availability of each generating unit and hourly load over specified simulation horizon are given. Calculate the production cost using STS formulation. Let L^i be the production cost of the i-th sample.
5. Calculate the residual term $U^i = L^i - Z^i + E\{Z\}$
6. Update the estimate of $E\{U\}$ and $V\{U\}$

$$E\{U\}=\frac{1}{NS}\sum_{i=1}^{NS}U^i \qquad \text{and} \qquad V\{U\}=\frac{1}{NS}\sum_{i=1}^{NS}\left(U^i-E\{U\}\right)^2$$

where *NS* is the number of samples. If the relative uncertainty is acceptable, stop; otherwise go to Step 2. We use relative uncertainty of 5% when the relative uncertainty is defined as:

$$\kappa=\frac{\left(V\{U\}/NS\right)^{0.5}}{E\{U\}}$$

5.6.2 Results. The minimization of production and maintenance costs is used as the objective function of LTS. We use the hierarchy described in Figure 5.1. The minimum reserve requirement for the GENCO is assumed 30 MW.

The results of the following test cases are included to show the effect of dynamic scheduling on maintenance schedule and reserve in the system.

- Case 1: Maintenance schedule is not updated based on STS results (MP1 only).
- Case 2: In addition of Case 1, dynamic maintenance schedule is incorporated (without SP1).
- Case 3: In addition of Case 2, ISO constraints are incorporated.

Basically, Case 1 solves the MP1 problem in which the maintenance problem is the master problem and fuel dispatch is the sub-problem. Using Benders decomposition, the LTS solution is converged in two iterations. The corresponding generator maintenance schedules for the weekly load are shown in Table 5.1.

Table 5.1 Generating Unit Maintenance Schedule for Case 1

Unit	Weeks on Maintenance
1	24,25
2	27,28
3	20,21
4	22,23
5	26,27

In Case 1, we have made similar calculations as in Case 0 of Section 3.5 of Chapter III. In Chapter III, we calculated the generating unit maintenance schedule without network constraints. Here, we add fuel and probabilistic constraints to Case 0 of Section 3.5. These constraints will limit the output of all generating units. Unit 3, which was always on in Case 0 of Chapter III, is forced to be off in weeks 22 and 23. Unit 5 is on more often to compensate for unit 3. The maintenance schedule of unit 5 is shifted from weeks 22-23 to weeks 26-27.

In Case 2, we apply dynamic scheduling. Given the maintenance schedule in MP1, we simulate the STS solution using the Monte-Carlo sampling to mimic the GENCO's realization. The sample which has the worst reserve case is seen in Figure 5.4. The reserve decreases below the minimum 30 MW limit because of STS constraints. The reserve between weeks 18 and 20 is above 30 MW because there is no generating unit on maintenance during these weeks.

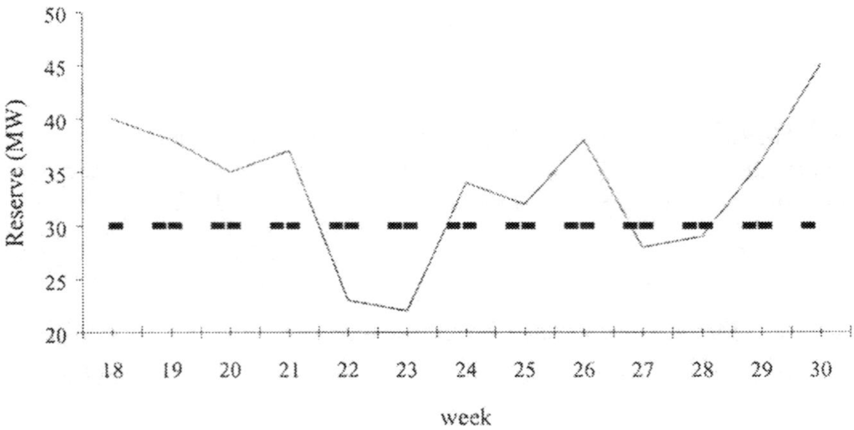

Figure 5.4 Generation Reserve without Dynamic Scheduling

Imposing dynamic scheduling, we show in Table 5.2 the re-scheduling stages of generating units during simulations. The final unit maintenance schedule is shown in Table 5.3.

Table 5.2 Unit Rescheduling during Simulations

Unit	Rescheduled Week			
	18	22	27	28
1	24,25	**23,24***	N/A	N/A
2	27,28	27,28	27,28	N/A
3	20,21	N/A	N/A	N/A
4	22,23	**24,25**	N/A	N/A
5	26,27	**27,28**	**28,29**	28,29

*In week 22, unit 1 is rescheduled for maintenance in weeks 23,24

The second column of Table 5.2 shows the maintenance schedule for units 1 to 5 in the beginning of week 18, which is the initial schedule of the study. Given this initial schedule, STS is simulated using the Monte Carlo sampling to mimic the actual system realization. Over a given time period, uncertainties will be resolved and new ones will evolve. To account for this, LTS will reschedule the generating units depending on the system performance and new pieces of information that may become available during the simulation. In week 22, it is found that the initial schedule needs to be modified since the system reserve is below 30 MW. The reserve is depicted in Figure 5.5. As a result, we recalculate the LTS in the beginning of week 22 to reschedule the generating unit maintenance for the rest of the simulation. This updated schedule is given in the third column of Table 5.2. Maintenance schedule of units 1, 4 and 5 which are shown in bold are changed accordingly. Unit 1 which was scheduled for maintenance in weeks 24, 25 is rescheduled for maintenance in weeks 23, 24. Units 4 and 5 are also rescheduled from weeks 22, 23 and 26, 27 into weeks 24, 25 and 27, 28. Unit 3 is not considered as part of the units that need maintenance since its maintenance has been done in weeks 18 and 19. As seen in Table 5.2, the unit maintenance is also rescheduled at the beginning of week 27 since the system reserve tends to drop below 30 MW (see Figure 5.5). In week 27, unit 5 is rescheduled. The peaking unit 5 is rescheduled more often because of its scheduling flexibility.

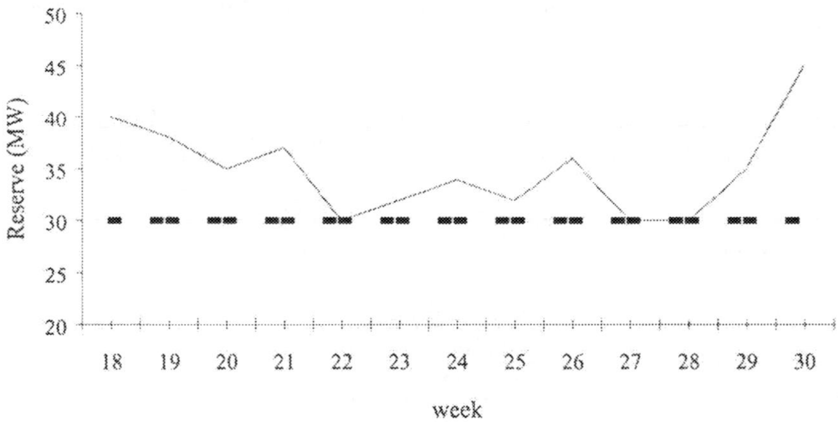

Figure 5.5 Generation Reserve with Dynamic Scheduling

Table 5.3 Generating Unit Maintenance Schedule for Case 2

Unit	Weeks on Maintenance
1	23,24
2	27,28
3	20,21
4	24,25
5	28,29

In Case 3, The ISO (network) sub-problem is imposed on the optimization problem. The schedules given in Table 5.3 are submitted to the ISO to be checked. The results indicate over-flow transmission violations in the system and the maximum flow violations over the three-month study period are shown in Table 5.4. Other line flows are within limits for most of the time periods.

Table 5.4 Overflow in Transmission Lines (12-week horizon, ϵ=1% of load)

Line Between buses	Rating MVA	Peak Flow MVA
1-3	208	231
11-14	600	605

These violations are signaled to the GENCO by sending Benders cuts (dual price with respect to the violations) to MP1. These constraints are added to the GENCO's formulation in MP1 to solve LTS and coordinate it with STS again. The dual price signal is a price which, if introduced into MP1, will result in a solution consistent with the original constraints under which the signal was derived. The iterative process is coordinated by estimating and refining a price signal to which each generator responds. The imposed ISO constraints increase the cost of maintenance and production. Table 5.5 shows the change in the total cost over the study period, indicating a more expensive maintenance schedule to satisfy network constraints. The final generating unit maintenance schedule for Case 3 is given in Table 5.6.

Table 5.5 Total Cost for Generating Unit (12-week horizon, ϵ=1% of load)

Case	Cost ($\$10^6$)
1	36.63702
2	37.02694
3	37.38971

Table 5.6 Generating Unit Maintenance Schedule for Case 3

Unit	Weeks on Maintenance
1	23,24
2	28,29
3	18,19
4	24,25
5	27,28

A change in reserve requirements will result in a different maintenance cost. In order to examine the cost difference, we conduct the Case 3 in a 52-week study horizon with different reserve limits. Increasing the reserve limit somewhat shifts the maintenance pattern as additional scheduled unit outages, due to maintenance, are scheduled in the summer peak load period and less in the fall load valley.

The relation between the reserve limit and cost are not linear as seen in Figure 5.6. The maintenance cost increases in stepwise. Increasing the reserve limit from 30MW to 40 MW does not change the maintenance schedule since the available units will pick up the increment. If we increase the required reserve capacity beyond 40 MW, the available units will no longer be able to pick up the increment. In this case, generating unit maintenance schedule will be shifted to satisfy the reserve requirement, which in turn will increase the maintenance cost. The increasing maintenance cost caused by increasing the required reserve capacity will reach a saturation level at the point where the generating unit maintenance constraints are bounded. This means that all units are scheduled for maintenance at their most expensive time.

The additional generation reserve capacity could be made available for sale in the power market. Figure 5.6 could be used, as a tool, by the GENCO to determine the reserve limit that would maximizes its revenue. Figure 5.6 shows that the maintenance cost does not change if the reserve increases from 60 MW to 70 MW. It implies that the GENCO could sell 70 MW of reserve capacity instead of 60 MW with no additional maintenance cost. The GENCO could also decide the cost/revenue benefits of providing 40 MW versus 70 MW reserve capacity given that the maintenance cost would increase from $\$5.43 \times 10^6$ to $\$5.45 \times 10^6$.

Here, we considered one GENCO in our case studies. However, in the case of multiple GENCOs, the cuts will be divided linearly among entities.

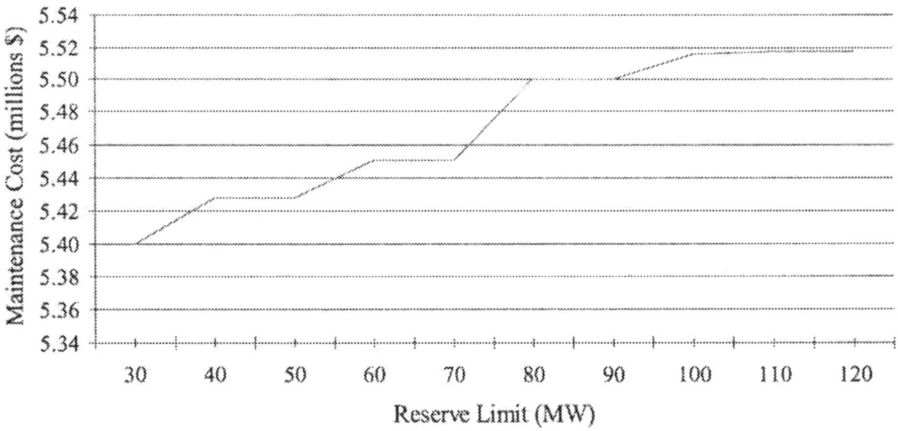

Figure 5.6 Effect of Reserve Limit on Maintenance Cost

CHAPTER VI

LONG-TERM TRANSMISSION MAINTENANCE SCHEDULING

6.1 INTRODUCTION

Most of the maintenance scheduling methods available in the literature consider only the generation system and do not take into account the maintenance of transmission lines. Transmission line maintenance scheduling ought to minimize the total maintenance cost and TRANSCOs' loss of revenue while satisfying the reliability and network requirements. It is essential for TRANSCOs to include forced outages of their components in transmission maintenance scheduling, which may otherwise result in unrealistic reliability indices. The proposed transmission maintenance schedule will be submitted by TRANSCOs to the ISO, which will check network flow constraints based on the generation information provided by GENCOs. The methodology for the solution of this long-term transmission maintenance problem is discussed in this chapter.

In considering network flow constraints, we deal with independent entities such as the ISO, GENCO and TRANSCO. Hence, coordination based on Benders decomposition will be the most logical solution to this maintenance problem.

In this chapter, we first present the formulation in a deterministic approach and proceed with the formulation of a probabilistic approach. Section 6.2 describes the formulation of the deterministic approach. The solution of the network model, the maintenance master problem (TRANSCO) and the operation sub-problem (ISO) are discussed in Section 6.3. The results for a simple example and the application of the proposed method to the IEEE-RTS (see Appendix A) are presented in Sections 6.4 and 6.5, respectively. The probabilistic formulation and its solution are discussed in Sections 6.6 to 6.8.

6.2 PROBLEM FORMULATION

Mathematically, the transmission maintenance problem is formulated as follows:

$$Min \sum_{t} \left(\sum_{k \in \omega} C'_{kt} \left(\overline{N}_k - N_{kt} \right) + \sum_{i \in \Omega} \{ c_{it} r_{it} \} \right)$$

S.T.

maintenance constraints :

$$N_{kt} = \overline{N}_k \qquad for\ t \le e'_k \quad or \quad t \ge l'_k + d'_k \qquad (i)$$

$$N_{kt} < \overline{N}_k \qquad for\ s'_k \le t \le s'_k + d'_k$$

$$N_{kt} \in \{0,1,..\overline{N}_k\} \quad for\ e'_k \le t \le l'_k$$

1. crew availability *3. seasonal limitations*

2. resources availability *4. desirable schedule* *(ii)*

network constraint s :

$$\boldsymbol{Sf} + \boldsymbol{r} = \boldsymbol{d} - \boldsymbol{g} \qquad \forall t \qquad (iii) \qquad (6.1)$$

$$|\boldsymbol{f}| \le \bar{\boldsymbol{f}}.\boldsymbol{N} \qquad \forall t \qquad (iv)$$

The objective seeks the minimum transmission line maintenance cost for the whole planning period. The first term in the objective function of (6.1) is the transmission line maintenance cost. The set of lines that require maintenance is ω. The second term is the expected loss of revenue due to line maintenance. The loss of revenue is proportional to the amount of scheduled energy that cannot be delivered to receiving buses due to line maintenance. The set of receiving buses is Ω.

Constraints (i) represent the weekly maintenance window stated in terms of maintenance variables N_{kt}. Lines must be available before their earliest possible period of maintenance (e'_k) and after their latest possible period of maintenance (e.g., $l'_k + d'_k$). We introduce variable y as a decision variable; its value is 0 if M_k lines in right-of-way k are off-line for maintenance otherwise its value is 1.

$$N_{k\tau} = (\overline{N}_k - M_k) + y_{kt} M_k$$

$$y_{kt} \in \{0,1\}$$

(6.2)

To avoid interruption during maintenance, maintenance constraints (i) is formulated as:

$$y_{kt} = 1 \qquad for\ t \le e'_k \quad or \quad t \ge l'_k + d'_k$$

$$(Y^{off}_{k(t-1)} - d'_k)(y_{kt} - y_{k(t-1)}) \ge 0 \qquad for\ e'_k \le t \le l'_k$$

$$\sum_t y_{kt} = l'_k - e'_k - d'_k \qquad for\ e'_k \le t \le l'_k$$

Additional constraints consisting of crews and other resources availability, seasonal limitations and favorable schedules can be incorporated into e'_k and l'_k of constraint (ii). If we consider, for example, that lines 1, 2 and 3 should be on maintenance simultaneously, the set of constraints is formed as follows:

$$y_{1t}+y_{2t}+y_{3t}=3 \qquad \text{or} \qquad y_{1t}+y_{2t}+y_{3t}=0$$

If we consider limited resources and crews available in each maintenance area, the set of constraints is formed as follows:

$$\sum_{k \in A} \sigma_{mk} M_k (1 - y_{kt}) \leq \psi_{mt}$$

In the case of limited resources, ψ_{mt} would represent the amount of resource m available in area A for each time t, and σ_{mk} would represent a percentage of this resource required for the maintenance of line k. In the case of crew constraints, ψ_{mt} would be the number of maintenance crews of type m in area A and σ_{mk} would be a percentage of these crews required for maintenance of line k.

The long-term maintenance scheduling solves a relaxed problem using a transportation model in which the Kirchoff laws are dropped. The solution will be forwarded to the ISO, which utilizes a transportation model to represent system operation limits. Set of constraints (iii) and (iv) represents peak load balance and other operational constraints such as power injections by the GENCO and transmission capacity limits.

The problem in (6.1) has a special structure that enables it to be decomposed into a maintenance master-problem, similar to Chapter II, and a subproblem.

6.3 SOLUTION METHODOLOGY

Let N denote the vector of maintenance variables $\{N_{kt}\}$ and ω represent the set of maintenance schedules for which constraints (iii)-(vii) are satisfied in all periods t. To ensure $N \in \omega$, the maintenance schedule must be feasible. The tth sub-problem is feasible if and only if an optimal value exist for the following problem:

$$Min \sum_{t} \sum_{i \in \Omega} (c_{it} r_{it})$$

$$S.T. \tag{6.3}$$

$$\mathbf{Sf} + \mathbf{r} = \mathbf{d} - \mathbf{g} \qquad \forall t \qquad (i)$$

$$|\mathbf{f}| \leq \bar{\mathbf{f}}.\mathbf{N} \qquad \forall t \qquad (ii)$$

The Lagrangian dual in this case is $max_{\nu,\lambda'\geq 0} U_t(\nu,\lambda')$ where $U_t(\nu,\lambda')$ is the following Lagrangian function and ν,λ' are multipliers of constraints (i)-(ii) in (6.3):

$$U_t(\nu,\lambda') = min_{|f|\geq 0}\left\{\begin{array}{l}\sum_{i\in\Omega} c_{it} r_{it} + \sum_i \nu_{it}\left(\left(\sum_{k\in i} S_{ik} f_{kt}\right) + g_{it} + r_{it} - d_{it}\right) + \\ + \sum_k \lambda'_{kt}\left(|f_{kt}| - \bar{f}_k\right) N_{kt}\end{array}\right\}$$

We then arrive at the following generalized Benders master problem:

$$Min \quad z$$

$$S.T.$$

$$z \geq \sum_t \left(\sum_k C'_{kt}(\overline{N}_k - N_{kt})\right)$$

$$\sum_t U_t(\nu,\lambda') \leq z \qquad \qquad for\ all \quad \nu,\lambda' \geq 0$$

maintenance constraints :

$$\begin{array}{lll} N_{kt} = \overline{N}_k & for\ t \leq e'_k\ \ or\ \ t \geq l'_k + d'_k & \\ N_{kt} < \overline{N}_k & for\ s'_k \leq t \leq s'_k + d'_k & (i)\quad (6.4) \\ N_{kt} = \{0,1,..,\overline{N}_k\}\ for\ e'_k \leq t \leq l'_k & & \end{array}$$

1. crew availability   *3. seasonal limitations*

2. resources availability   *4. desirable schedule*   *(ii)*

In this study, the initial maintenance master problem is formulated as follow:

$$Min \ z$$

$$S.T. \quad z \geq \sum_t \left\{\sum_k C'_{kt}\ \bar{f}_k\ (\overline{N}_k - N_{kt})\right\}$$

maintenance constraints :

$$N_{kt} = \overline{N}_k \qquad for\, t \le e'_k \quad or \quad t \ge l'_k + d'_k$$

$$N_{kt} < \overline{N}_k \qquad for\;\; s'_i \le t \le s'_i + d'_i \tag{6.5}$$

$$N_{kt} \in Integer$$

1. crew availability *3. seasonal limitations*

2. resources availability *4. desirable schedule*

Now we consider the subproblem. The Benders cut for the nth iteration of the ISO sub-problem resulting from the nth solution of the master problem is as follows:

$$\sum_{i \in \Omega} c_{it} r_{it}^{n} + \sum_{k} \lambda'^{n}_{kt} \; \bar{f}_k \, (N_{kt}^{n} - N_{kt}) \le z \tag{6.6}$$

Similar to (2-20) in Chapter II, the multiplier $\lambda'^{\,n}_{kt}$ may be interpreted as the marginal decrease in the loss of revenue with a 1 MW increase in transmission lines' available capacity, given the nth iteration of maintenance schedule. At each iteration, the solution of the set of ISO sub-problems will result in a set of one or more cuts. The cuts in (6.6) will force the maintenance value, N_{kt}, to reduce the loss of revenue. The new cuts (constraints) will be added to the relaxed master problem, which will then be solved in the next iteration.

The revised TRANSCO's master problem that consists of cuts is as follows:

$Min\;\; z$

$S.T.$

$$z \ge \sum_{t} \sum_{k} \left\{ C'_{kt} \bar{f}_k \, (\overline{N}_k - N_{kt}) \right\}$$

•) *maintenance constraint :*

$$N_{kt} = \overline{N}_k \qquad for\, t \le e'_k \quad or \quad t \ge l'_k + d'_k$$

$$N_{kt} < \overline{N}_k \qquad for\;\; s'_i \le t \le s'_i + d'_i \tag{6.7}$$

1. crew availability *3. seasonal limitations*

2. resources availability *4. desirable schedule* *5.etc.*

•) *additional cuts are :*

$$\sum_{i} c_{it} r_{it}^{n} + \sum_{k} \left(\lambda'^{\,n}_{kt} \; \bar{f}_k \, (N_{kt}^{n} - N_{kt}) \right) \le z$$

$$N \le \overline{N}$$

where,

n is the current number of iteration

λ'^{n}_{kt} is the multiplier vectors at the nth iteration

6.4 EXAMPLE

We use the same three-bus system as in Section 2.5. The potential loss of revenue at each bus due to line maintenance is $250/p.u. In addition to the data given in Section 2.5, transmission line data in per unit are given in Table 6.1. The problem is defined as performing maintenance for a minimum of two lines. We assume that the study period merely has one time interval in which loads are assumed constant.

Table 6.1 Line Maintenance Data

Line	Ω/line	# of lines	Maint.Cost/line ($/line)	Capacity/line (p.u.)
1-2	0.2	2	200	0.5
2-3	0.25	2	300	0.5
1-3	0.4	2	100	0.5

First, we solve the initial master problem for TRANSCO.

Maintenance master problem iteration 1:

Min z

S.T. $200*(2-N_{12})+300*(2-N_{23})+100*(2-N_{13}) \le z$

$N_{12}+N_{23}+N_{13} \le 4$

$N_{12}\le 2 \qquad N_{23}\le 2 \qquad N_{13}\le 2$

$N_{12}\in\{0,2\} \qquad N_{23}\in\{0,2\} \qquad N_{13}\in\{0,2\}$

The solution is: $N_{12}=2 \quad N_{23}=2 \quad N_{13}=0 \quad z=200.$

The ISO sub-problem iteration 1:

The presumption is that the ISO knows the GENCO's schedule: $g_1 = 0$ p.u., $g_2 = 2$ p.u. and $g_3 = 3$ p.u. The ISO solves the sub-problem for the first trial of the maintenance schedule as follows:

Min $250\,(r_1+r_2+r_3)$

S.T. $-f_{12}-f_{13}+r_1 = 1-0$ Load balance at bus 1

$-f_{23}+f_{12}+r_2 = 3-2$ Load balance at bus 2

$f_{13}+f_{23}+r_3 = 1-3$ Load balance at bus 3

$-2*0.5 \le f_{12} \le 2*0.5$ Line 1-2 flow limit

$-2*0.5 \le f_{23} \le 2*0.5$ Line 2-3 flow limit

$0.0 \le f_{13} \le 0.0$ Line 1-3 flow limit

The primal solution of the operation sub-problem is: $r_1=0$, $r_2=1.0$, $r_3=0$.
The dual price of the operation sub-problem is: $\lambda'_{12}=0$, $\lambda'_{23}=250$, $\lambda'_{13}=250$.

The loss of revenue cut is as follows: $250.0+ 250*0.5\ (2- N_{23}) +250*0.5\ (0- N_{13}) \leq z$

Maintenance master problem iteration 2:

$$
\begin{aligned}
&Min\ \ z\\
&S.T.\ 200*(2- N_{12})+ 300*(2-N_{23})+ 100*(2-N_{13}) \leq z\\
&\quad 250 + 250*0.5\ (0- N_{12}) +250*0.5\ (2- N_{13}) \leq z\\
&\quad N_{12}+N_{23}+N_{13} \leq 4\\
&\quad N_{12}\leq 2 \qquad N_{23}\leq 2 \qquad N_{13}\leq 2\\
&\quad N_{12}\in\{0,2\} \qquad N_{23}\in\{0,2\} \qquad N_{13}\in\{0,2\}
\end{aligned}
$$

The solution is: $N_{12}= 0$, $N_{23}=2$, $N_{13}=2$, $z=400$.

The ISO sub-problem iteration 2:

Given the trial maintenance schedule in iteration 2, the dual prices in ISO's sub-problem of iteration 2 are zero. So, we stop here as the solution has converged.

6.5 CASE STUDIES

We apply the proposed method to the system in Section 2.6. Given the generation maintenance data by the GENCO, transmission facilities maintenance data proposed by the TRANSCO for a particular area are checked by the ISO within the study period. During the three-month period, at least 4 transmission lines in the maintenance area will require maintenance. Table 6.2 gives the transmission data within the maintenance area. The maintenance cost for all lines is assumed $\$0.72\text{x}10^4$ /mile and the potential loss of revenue at each bus due to line maintenance is \$250/p.u. Detailed system data for transmission lines, generators and loads are given in Appendix A.

Table 6.2 Transmission Line Data

From Bus	To Bus	Length (Miles)	Maintenance		Rating (MVA)
			Window (weeks)	Duration (weeks)	
1	2	3	18-29	1	193
1	3	55	18-29	1	208
1	5	22	18-29	1	208
2	4	33	18-29	1	208
2	6	50	18-29	1	208
3	9	31	18-29	1	208
4	9	27	18-29	1	208
5	10	23	18-29	1	208
6	10	16	18-29	1	193
7	8	16	18-29	1	208
8	9	43	18-29	1	208
8	10	43	18-29	1	208

The minimization of maintenance costs is used as the objective function. The "Energy not served" in each week is limited to a maximum of 1% of the total energy. The results of the following test cases will show the effect of network constraints on maintenance schedule.

- Case 1: We consider transmission maintenance without ISO's network flow constraints.
- Case 2: The ISO will check the proposed transmission line maintenance schedule.

In Case 1, maximum transmission flows over the three-month study period are shown in Table 6.3. The corresponding line maintenance schedule is shown in Tables 6.4. Case 2 studies the effect of the ISO's subproblem on maintenance schedule in which transmission limits are imposed on the optimization problem. The imposed transmission limits increase the cost of maintenance. Table 6.5 shows the change in maintenance cost over the study period as we introduce network flow constraints, indicating a shift from inexpensive maintenance costs to those with more expensive maintenance costs.

Here, we considered one TRANSCO in the case studies. However, in the case of multiple TRANSCOs, the cuts will be divided linearly among entities.

Table 6.3 Overflow on Transmission Lines (12-week horizon)

Case 1		
Line Between Buses	MW Rating	Peak MW Flow
1-3	208	221
1-2	193	212
1-5	208	233

Table 6.4 Effect of Transmission Capacity on Transmission Maintenance Schedule (12-week horizon)

Case	Line Between Buses	Week on Maintenance
1	1-2	20
	5-10	21
	4-9	22
	6-10	26
2	2-4	20
	5-10	25
	4-9	27
	3-9	26

Table 6.5 Total Cost for Maintenance (12-week horizon)

Case	Cost (10^6 \$)
1	0.4968
2	0.9006

6.6 PROBABILISTIC FORMULATION

The need to represent operating uncertainties in planning problems is widely recognized. In our study, maintenance scheduling evaluates the availability of generating units and transmission lines in the planning period, taking into account equipment outages and load variations (i.e., composite reliability evaluation).

Again, we use a transportation model for the ISO to represent system operation limits, peak load balance equation, and generation and line capacity limits. The formulation that considers the outages of equipment is formulated as follows:

$$Min \sum_t \left(\sum_{k \in \omega} C'_{kt} (\overline{N}_k - N_{kt}) + \sum_{i \in \Omega} E\{c_{it} r_{it}\} \right)$$

$S.T.$

maintenance constraints :

$$N_{kt} = \overline{N}_k \qquad for\ t \le e'_k \quad or \quad t \ge l'_k + d'_k \qquad (i)$$

$$N_{kt} < \overline{N}_k \qquad for\ s'_k \le t \le s'_k + d'_k$$

$$N_{kt} \in \{0,1,..\overline{N}_k\}\ for\ e'_k \le t \le l'_k$$

1. crew availability *3. seasonal limitations*

2. resources availability *4. desirable schedule* *(ii)*

network constraints :

$$\boldsymbol{Sf} + \boldsymbol{r} = \boldsymbol{d}(\varphi) - \boldsymbol{g}(\varphi) \qquad \forall t \qquad (iii)\ (6.8)$$

$$|\boldsymbol{f}| \le \bar{\boldsymbol{f}}(\varphi).\boldsymbol{N} \qquad \forall t \qquad (iv)$$

Hence, the initial maintenance master problem is formulated as follow:

$Min \quad z$

$$S.T. \quad z \ge \sum_t \sum_k \left\{ C'_{kt} (\overline{N}_k - N_{kt}) \right.$$

maintenance constraints :

$$
\begin{aligned}
&N_{kt} = \overline{N}_k && \text{for } t \le e'_k \text{ or } t \ge l'_k + d'_k \\
&N_{kt} < \overline{N}_k && \text{for } s'_i \le t \le s'_i + d'_i \\
&N_{kt} \in \{0,1,\ldots,\overline{N}_k\} && \text{for } e'_i \le t \le l'_i
\end{aligned} \tag{6.9}
$$

1. crew availability *3. seasonal limitations*

2. resources availability *4. desirable schedule*

And, the ISO sub-problem is given as:

$$
\begin{aligned}
&Min\ E\left\{\sum_{i \in \Omega} c_{it} r_{it}\right\} \\
&S.T. \\
&\quad \boldsymbol{Sf + r = d(\varphi) - g(\varphi)} \\
&\quad |\boldsymbol{f}| \le \bar{\boldsymbol{f}}(\boldsymbol{\varphi}).\boldsymbol{N}^n \qquad (dual\ variable\ is\ \lambda')
\end{aligned} \tag{6.10}
$$

For each ISO sub-problem, resulting from the nth solution of the TRANSCO (master problem), the Benders cut is associated with a deterministic, instead of a probabilistic, sub-problem. The cut is

$$
E\left\{\sum_{i \in \Omega} c_{it} r_{it}^n\right\} + \sum_k \lambda_{kt}^n\ \bar{f}_k\,(N_{kt}^n - N_{kt}) \le z \tag{6.11}
$$

The solution of (6.10) involves the reliability evaluation of a composite system. The procedure is as follows:

1. Select a system state φ, i.e., define load levels, equipment availability, operating conditions, etc.
2. Calculate (6.10) for the selected state, i.e., verify whether that specific configuration of transmission lines is able to supply that specific load without violating system limits.
3. Update the estimated load shedding expectation.
4. Return to step 1 if the accuracy is unacceptable.

In order to allow a fast algorithm and efficient calculations, we use a simultaneous decomposition simulation for all generator and transmission states and intervals [Den92].

Hence, the revised master problem that consists of feasible and infeasible cuts is as follows:

$$Min \quad z$$

$$S.T. \quad z \geq \sum_t \sum_k \left\{ C'_{kt} (\overline{N}_k - N_{kt}) \right\}$$

• *maintenance constraints :*

$$N_{kt} = \overline{N}_k \qquad for\; t \leq e'_k \quad or \quad t \geq l'_k + d'_k$$

$$N_{kt} < \overline{N}_k \qquad for\; s'_i \leq t \leq s'_i + d'_i$$

$$N_{kt} \in \left\{0,1,\ldots,\overline{N}_k\right\} \; for\; e'_i \leq t \leq l'_i$$

1. crew availability   *3. seasonal limitations*

2. resources availability   *4. desirable schedule*

• *one or more subproblems cuts as :*

$$E\left\{ \sum_{i \in \Omega} r_{it}^n \right\} + \sum_k \lambda'^{n}_{kt} \; \bar{f}_k \, (N_{kt}^n - N_{kt}) \leq z \qquad (6.12)$$

$\forall t \in$ *infeasible subproblem*

$N \leq \overline{N}$

where: n is the current number of iterations,

λ'^n is the multiplier vector at the nth iteration.

6.7 EXAMPLE

Since we are considering the transmission maintenance scheduling problem, the forced outrage rate (FOR) of each line is given in Table 6.6. The potential loss of revenue at each bus due to line maintenance is \$250/p.u. The problem considers the maintenance of at least 2 lines. We assume the study period has only one time interval and loads are assumed constant during the study period.

Table 6.6 Line Forced Outage Rate

Line	Max Capacity (p.u.)	FOR
1-2	1.0	0.05
2-3	1.0	0.1
1-3	1.0	0.02

First, we solve the initial maintenance master problem:

Min z

S.T. $200*(2-N_{12})+300*(2-N_{23})+100*(2-N_{13}) \le z$

$N_{12}+N_{23}+N_{13} \le 4$

$N_{12}\le 2 \qquad N_{23}\le 2 \qquad N_{13}\le 2$

$N_{12}\in\{0,2\} \qquad N_{23}\in\{0,2\} \qquad N_{13}\in\{0,2\}$

The solution is: $N_{12}=2$ $N_{23}=2$ $N_{13}=0$ $z=200$

The sub-problem given the first trial of maintenance schedule is as follows:

Min $250\,E(r_1+r_2+r_3)$

S.T. $-f_{12}-f_{13}+r_1 = 1-0$	Load balance at bus 1
$-f_{23}+f_{12}+r_2 = 3-2$	Load balance at bus 2
$f_{13}+f_{23}+r_3 = 1-3$	Load balance at bus 3
$-2*0.5 \le f_{12} \le 2*0.5$	Line 1-2 flow limit
$-2*0.5 \le f_{23} \le 2*0.5$	Line 2-3 flow limit
$0.0 \le f_{13} \le 0.0$	Line 1-3 flow limit

The primal solutions for all state spaces in the feasibility check are given in Table 6.7.

Table 6.7 Feasibility Check State Spaces

φ		Prob.	r	Prob. x r
f_{12}	f_{23}			
1	1	0.855	1.0	0.855
0	1	0.045	1.0	0.045
1	0	0.095	2	0.190
0	0	0.005	2	0.010
				$E\{r\}$=1.100

The dual prices of the sub-problem are calculated by increasing the line capacity by 1 MW. If we increase the capacity of Line 2-3 to 2 MW, we get $E\{r\}=0.245$. So, we can calculate the dual price of Line 2-3 as $\lambda_{f23}=(1.100-0.245)*250=213.75$

Using the same process for line 1-2 and 1-3, we get dual prices of each line as:

$\lambda_{f12}=0 \qquad \lambda_{f23}=213.75 \qquad \lambda_{f13}=245.0$

The cut is as follows: $275+213.75*0.5\,(2-N_{23})+245.0*0.5\,(0-N_{13}) \le z$

The solution of the maintenance master problem for the second iteration is:

$N_{12}=0$ $N_{23}=2$ $N_{13}=2$ $z=400$

Given the trial maintenance schedule from iteration 2, we calculate the dual price of the sub-problem and continue until the value of z does not change any more.

6.8 CASE STUDIES

In Case 3, we apply the proposed method to the system in Section 6.5 except now we include transmission outages as well. The minimization of production and maintenance costs is used as the objective function. Expected Energy Not Supplied (ε) in each week is limited to the maximum of 1% of the total energy. The results of the following test case are included to show the effect of probabilistic network constraints on maintenance schedule.

The final maintenance costs are given in Table 6.8 which again shows the additional maintenance cost for the probabilistic formulation. Here, if we implement the schedule of Case 2 using the probabilistic approach, we detect additional line flow violations as shown in Table 6.9. Once we remove the violations, the corresponding line maintenance schedule of Case 3 is shown in Table 6.10. We will discuss the coordination between long-term and short-term transmission line maintenance scheduling problems in Chapter 7.

Table 6.8 Total Cost for Maintenance (12-week horizon, ε=1% of load)

Case	Cost (10^6 \$)
1	0.4968
2	0.9006
3	0.9211

Table 6.9 Overflow on Transmission Lines (12-week horizon, ε=1% of load)

Case 2		
Line between buses	MW Rating	Peak MW flow
1-3	208	223
1-2	193	215
1-5	208	235

Table 6.10 Effect of Transmission Capacity on Transmission Maintenance Schedule (12-week horizon, ε=1% of load)

Case	Line Between Buses	Week on Maintenance
1	1-2	20
	5-10	21
	4-9	22
	6-10	26
2	2-4	20
	5-10	25
	4-9	27
	3-9	26
3	2-4	21
	5-10	22
	4-9	27
	3-9	26

CHAPTER VII

COORDINATION BETWEEN LONG-TERM AND SHORT-TERM TRANSMISSION MAINTENANCE SCHEDULING

7.1 INTRODUCTION

The rapidly changing business environment for electric power utilities has resulted in unbundling of services provided by these utilities. Wheeling of electrical energy (transmission services) is one of the more prevalent unbundled services. The changing structure of electric power utilities and the emergence of independent power producers, independent customers and transmission provider are bringing transmission management into a new focus.

The United States Federal Energy Regulatory Commission (FERC) addressed some of the complexities surrounding transmission issues. FERC proposed a system of point-to-point transmission capacity reservations under Capacity Reservation Open Access Transmission Tariffs (CRT) [Fed96]. The basic concept of CRT was that all transmission services would be reserved, and all reserved services would be firm services. According to CRT, transmission providers could offer two types of transmission reservations, i.e., firm and non-firm reservations. In firm reservations, transmission curtailment would not occur for economic reasons. Transmission services would be curtailed only when the system reliability was threatened or in emergency conditions. Firm transmission services were mostly provided to native loads and contracted wheeling transactions. In non-firm reservations, transmission providers would have the right to interrupt all or part of transmission services for various reasons including economics. Transmission providers could offer non-firm transmission services based on the capacity that was not used by customers with reservations, or from the capacity that was not previously reserved by customers. The FERC paper dealt with political and economic issues in detail but did not cover engineering issues. The cost of operating, maintaining and expanding the transmission system that concerned transmission services was addressed in [Shi91, Shi96, Yu97]. Most others addressed rate designs for wheeling services [Shi89, Hap94].

In an increasingly competitive environment, the risk of lost transmission sales must be considered seriously. A common integrated planning decision that confronts transmission providers is the line maintenance scheduling, since line maintenance involves a possibility of revenue loss. Line maintenance could mean that the transmission providers would be unable to sell as much services as they would have if all lines were fully operational. Scheduling line maintenance during low wheeling periods would be a smart move, but might also mean paying extra compensation to contractors and suppliers to work overtime. Transmission providers also have to maintain the quality of service and the reliability of the system when transmission lines are on maintenance. Engineering issues such as the bus voltages profile and transmission flow limits may need to be considered as well.

Transmission maintenance scheduling is an optimization problem with complex constraints. The maintenance schedule may span over several time periods, which may also impact the reliability of the system. In this connection, the following functions (Figure 7.1) are identified:

1. Long-term Transmission Maintenance Scheduling

   The long-term period (one year) is divided into intervals (weeks) and a maintenance scheduling strategy for the intervals is derived to satisfy weekly line maintenance constraints as well as seasonal constraints.

2. Power Transaction and Transmission Contracts

   During the power transaction process, GENCOs sell their energy to retailers, direct-access customers through bilateral contracts, and the Power Exchange. Transmission line contracts, such as network services that are based on usage time, or path services that are based on reservations, may follow after the power transaction process. The resulting information on generation schedules, load and line contracts will be used in short-term transmission maintenance scheduling.

3. Short-term Transmission Maintenance Scheduling

   If we assume that the line maintenance can be done in several hourly stages, we can determine the most appropriate hours within a given week when the short-term line maintenance can be done. According to the maintenance windows calculated in Step 1, and generation schedules and line contracts in Step 2, the hourly line maintenance is calculated to maximize the transmission providers' revenues while satisfying the system constraints and reliability requirements.

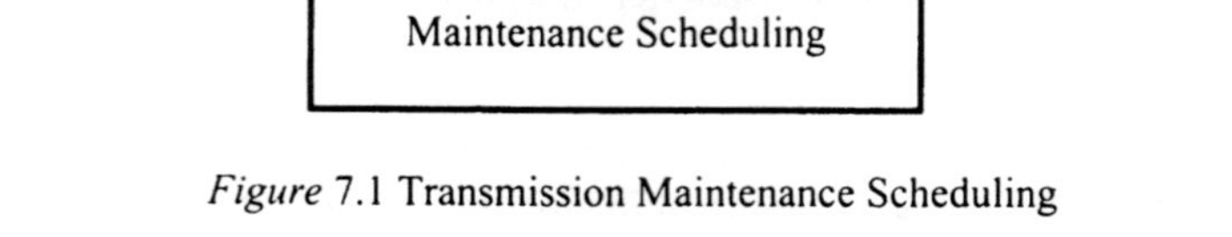

Figure 7.1 Transmission Maintenance Scheduling

In this chapter, we focus on the coordination between long-term and short-term transmission maintenance scheduling. We solve the problem of transmission line maintenance scheduling in order to maximize transmission providers' revenues. Our emphasis on transmission maintenance in this chapter is to take into account tradeoffs and technological constraints that impact transmission line maintenance scheduling. The "loss of revenue" introduced as an objective function in this chapter will be flexible enough to accommodate various pricing methods in different markets.

7.2 PROBLEM FORMULATION

Time horizons and increments represent differences between short-term and long-term maintenance scheduling. In long-term scheduling, the horizon of study is within one or two years with weekly increments (Δt), however, in short-term maintenance scheduling, the horizon of study is a few weeks with hourly increments ($\Delta\tau$). Long-term maintenance scheduling will determine specific windows for maintenance, and short-term maintenance scheduling will determine proper maintenance hours within that window.

7.2.1 Long-term Maintenance Scheduling.

The need to represent operating uncertainties in system planning is widely recognized. Long-term maintenance scheduling, which evaluates planned outages of generating units and lines in the planning period, ought to take into account equipment forced outages and load variations. As discussed in Chapter 6, the solution of the long-term maintenance will be forwarded to the ISO for checking network flow constraints according to GENCOs' data for generating units' availability. The ISO will use a transportation model to represent system operating limits, peak load balance, generation and line capacity limits, and generation and transmission outages (composite reliability evaluation). Long-term transmission maintenance was formulated in Chapter 6 as follows:

$$Min \quad \sum_t \left(\sum_{k\in\omega} C'_{kt}\,(\overline{N}_k - N_{kt}) + \sum_{i\in\Omega} E\{c_{it} r_{it}\} \right)$$

$St.$

maintenance constraints :

$$N_{kt} = \overline{N}_k \qquad for\ t \le e'_k \quad or \quad t \ge l'_k + d'_k \qquad (i)$$

$$N_{kt} < \overline{N}_k \qquad for\ s'_k \le t \le s'_k + d'_k$$

$$N_{kt} \in \{0, 1, .. \overline{N}_k\} \quad for\ e'_k \le t \le l'_k$$

1. crew availability   *3. seasonal limitations*

2. resources availability   *4. desirable schedule*   *(ii)*

network constraints :

$$\boldsymbol{Sf} + \boldsymbol{r} = \boldsymbol{d}(\varphi) - \boldsymbol{g}(\varphi) \qquad \forall t \qquad (iii)$$

$$|\boldsymbol{f}| \le \bar{\boldsymbol{f}}(\varphi).\boldsymbol{N} \qquad \forall t \qquad (iv)$$

(7.1)

7.2.2 Short-term Maintenance Scheduling.

The objective of short-term transmission maintenance scheduling is to minimize the maintenance cost and the loss of transmission revenues due to maintenance while satisfying maintenance requirements. The objective in this case is formulated as follows:

$$Min \sum_{\tau}\sum_{k\in\omega}\left\{C'_{k\tau}(\overline{N}_k - N_{k\tau})\right\}+\sum_{\tau}\sum_{i\in\Omega}\left\{c_{i\tau}r_{i\tau}\right\}+\sum_{\tau}\sum_{i\in\beta}\left\{c'_{i\tau}s_{i\tau}\right\} \tag{7.2}$$

The first term in (7.2) is the transmission line maintenance cost. The second term is the loss of revenue due to line maintenance. The loss of revenue is proportional to the amount of scheduled energy that cannot be delivered to receiving buses due to line maintenance. This implies that one or more contracts that are recallable may need to be canceled. Additional reactive power may also be needed to support the voltage profile due to line maintenance. Transmission providers will have two options for meeting the required reactive power generation. The first option is to interrupt all or parts of cancelable contracts. The second option is to purchase reactive power (ancillary services). The cost of these two options is represented by the third term of objective (7.2). The set of buses where there are ancillary services or cancelable reactive power contracts is β.

The sets of constraints are given as follows.

a) Maintenance constraints

$$\begin{aligned}
&y_{k\tau} = 1 && for\ y_{kt} = 1 \quad \forall\tau\in t\\
&(Y^{off}_{k(\tau-1)} - h_k)(y_{k\tau} - y_{k(\tau-1)}) \geq 0 && for\ y_{kt} = 0 \quad \forall\tau\in t\\
&\sum_{\tau} y_{kt} = n_\tau - h_k && for\ y_{kt} = 0 \quad \forall\tau\in t
\end{aligned} \tag{7.3}$$

Constraints (7.3) represent the hourly maintenance window stated in terms of maintenance variable $y_{k\tau}$. Here t is a master problem variable (TRANSCO) and τ is the subproblem variable (ISO). The maintenance window was determined by the long-term scheduling problem. In this regard, lines must be available when they are not scheduled to be on maintenance by long-term scheduling. Furthermore, short-term maintenance hours should be within the week that is scheduled by long-term.

b) Network flow constraints (ISO)

In the short-term horizon, we consider real power flow and voltage limits as network constraints. Network constraints will be considered by the ISO and could be decomposed into two smaller sets. The first set is for real power flow constraints, which limits transmission security violations based on the worst contingency case. The second set is for reactive power constraints, which limits voltage violations. The two sets of constraints are represented as follows:

- Limits on real power flow

Since we are primarily interested in screening transmission violations, we use the dc load flow solution to expedite the process. In this model, real power flows obey the two Kirchoff laws (nodal and loop). The real power flow between buses i and j is calculated as:

$$f_k = (\theta_i - \theta_j)\,\gamma_k \tag{7.4}$$

If we assume, $$|f_k| \le \bar{f}_k \tag{7.5}$$

Then, $$|\theta_i - \theta_j| \le \bar{\psi}_k \tag{7.6}$$

where $\bar{\psi}_k = \bar{f}_k / \gamma_k$.

Network constraints in matrix term are given as follows.

$$\begin{aligned} \boldsymbol{B}\boldsymbol{\theta} &= \boldsymbol{d} - \boldsymbol{g} \\ |\boldsymbol{S}'\boldsymbol{\theta}| &\le \bar{\boldsymbol{\Psi}} \end{aligned} \tag{7.7}$$

Let $\boldsymbol{P}=\boldsymbol{d}-\boldsymbol{g}$, then (7.7) becomes:

$$\begin{aligned} \boldsymbol{B}\boldsymbol{\theta} &= \boldsymbol{P} \\ |\boldsymbol{S}'\boldsymbol{\theta}| &\le \bar{\boldsymbol{\Psi}} \end{aligned} \tag{7.8}$$

Set of constraint (7.8) represents the load balance and other operational constraints such as transmission capacity limits.

- Limits on voltages

Transmission providers attempt to maintain a high quality of service when transmission lines are on maintenance. In the case of violations, this goal can be accomplished by canceling a few reactive power transactions, purchasing additional reactive power and adjusting tap-changing transformers. The reactive power constraints are formulated as follows:

- Reactive power operating reserve requirement

$$\sum_i \bar{Q}_i \ge Q_D \tag{7.9}$$

- Reactive power generation limits and load bus balance

$$\begin{aligned} \underline{\boldsymbol{Q}}_G \le \boldsymbol{Q}_G = \boldsymbol{F}_1(\boldsymbol{V}) &\le \bar{\boldsymbol{Q}}_G \\ \boldsymbol{Q}_D &= \boldsymbol{F}_2(\boldsymbol{V}) \end{aligned} \tag{7.10}$$

- System voltage and transformer tap limits

$$\underline{V} \le V \le \overline{V}$$
$$\underline{T} \le T \le \overline{T} \qquad (7.11)$$

We assume V^0 is the system voltage vector based on the initial unit commitment state. If V^0 is infeasible, additional adjustments to the generation and/or distribution of reactive power may be necessary in order to shift the system voltage from V^0 to a desired value V^*. We neglect the effect of reactive power adjustments on bus voltage angles. Based on the discussion given in Section 2.6, we linearized (7.11) in the vicinity of the initial operating point V^0 as given in (7.12).

$$Q_G^0 + J_1'' \Delta V \le \overline{Q}_G + \Delta \overline{Q}_T$$
$$J_2'' \Delta V = 0$$
$$\Delta \underline{V} \le \Delta V \le \Delta \overline{V} \qquad (7.12)$$

where $\Delta V = V^* - V^0$ is the incremental system voltage corresponding to ΔQ_G. J_1'' is the modified Jacobian matrix for buses connected to generators and transformers, and J_2'' is for load buses.

c) Line Reservation and Ancillary Services

To formulate firm and non-firm reservations, we include them in transmission flow constraints and the objective function. The firm reservation on line k at hour τ is represented as the minimum flow constraint $\underline{f_{k\tau}}$. Hence,

$$\underline{\Psi} \le |S'\theta| \le \overline{\Psi} \qquad (7.13)$$

To accommodate the flexibility of canceling non-firm reservation, we introduce slack variable r (load curtailment) into (7.8). Load curtailment, $r_{i\tau}$, should be smaller than or equal to the recallable load at bus i at hour τ. The active power constraints become:

$$B\theta + r = P$$
$$r \le \overline{r} \qquad (7.14)$$
$$\underline{\Psi} \le |S'\theta| \le \overline{\Psi}$$

Slack variables $r_{i\tau}$ is applied to buses where there are recallable transactions. Since additional reactive power may be needed to support the voltage profile, we introduce a slack variable s as an additional injected reactive power and let $\overline{Q} = \overline{Q}_G + \Delta\overline{Q}_T + Q_G^0$, then (7.12) becomes:

$$
\begin{aligned}
&J_1'' \, \Delta V - s \le \overline{Q} \\
&J_2'' \, \Delta V = 0 \\
&\Delta \underline{V} \le \Delta V \le \Delta \overline{V} \\
&s \le \overline{s}
\end{aligned}
\tag{7.15}
$$

Slack variables $s_{i\tau}$ apply only to the buses where there are recallable load or ancillary services. Canceling transaction or purchasing reactive power is related to the loss of revenue.

d) Overall Transmission Maintenance Formulation

The trade-off between maintenance cost and revenue loss is optimized in the proposed method. Mathematically, the short-term transmission line scheduling is stated in (7.16):

$$
Min \sum_{\tau} \sum_{k \in \omega} \left\{ C'_{k\tau} (\overline{N}_k - N_{k\tau}) \right\} + \sum_{\tau} \sum_{i \in \Omega} \left\{ c_{i\tau} r_{i\tau} \right\} + \sum_{\tau} \sum_{i \in \beta} \left\{ c'_{i\tau} s_{i\tau} \right\}
$$

St. *maintenance constraints*:

$$
\begin{aligned}
&y_{k\tau} = 1 && \text{for } y_{kt} = 1 \quad \forall \tau \in t \\
&(Y^{off}_{k(\tau-1)} - h_k)(y_{k\tau} - y_{k(\tau-1)}) \ge 0 && \text{for } y_{kt} = 0 \quad \forall \tau \in t \\
&\sum_{\tau} y_{kt} = n_\tau - h_k && \text{for } y_{kt} = 0 \quad \forall \tau \in t
\end{aligned}
$$

real power constraints:

$$
\begin{aligned}
&\boldsymbol{B\theta} + \boldsymbol{r} = \boldsymbol{P} \\
&\boldsymbol{r} \le \overline{\boldsymbol{r}} && \text{for } \forall \tau \\
&\underline{\boldsymbol{\Psi}} \le \left| \boldsymbol{S}' \boldsymbol{\theta} \right| \le \overline{\boldsymbol{\Psi}}
\end{aligned}
$$

reactive power constraints:

$$
\begin{aligned}
&J_1'' \, \Delta V - s \le \overline{Q} \\
&J_2'' \, \Delta V = 0 && \text{for } \forall \tau \\
&\Delta \underline{V} \le \Delta V \le \Delta \overline{V} \\
&s \le \overline{s}
\end{aligned}
\tag{7.16}
$$

7.3 SOLUTION METHODOLOGY

We solve long-term and short-term transmission maintenance scheduling problems independently. In each maintenance problem, the problem is decomposed into a master problem (TRANSCO) and operation subproblems (ISO). The master problem, which is an integer programming problem, is solved to generate a trial solution for maintenance schedule decision variables. This master problem is a relaxation of the original problem in that it contains only a subset of constraints. Its optimal value is a lower bound on the optimal value of the original problem. Once maintenance variables are fixed, the resulting operation subproblems will be treated as a set of independent subproblems, one for each time period, since there is no constraint across time periods. The set of operation subproblems are then solved by the ISO using the fixed maintenance schedule obtained from the solution of the master problem. At each iteration, the solution of subproblems generates dual multipliers which measure changes in either the production cost or the reliability resulting from marginal changes in the maintenance schedule. These dual multipliers are used to form one or more constraints (known as Benders cuts) which are added to the master problem for the next iteration. The process continues until a feasible solution is found whose cost is sufficiently close to the lower bound, as shown in Figure 7.2.

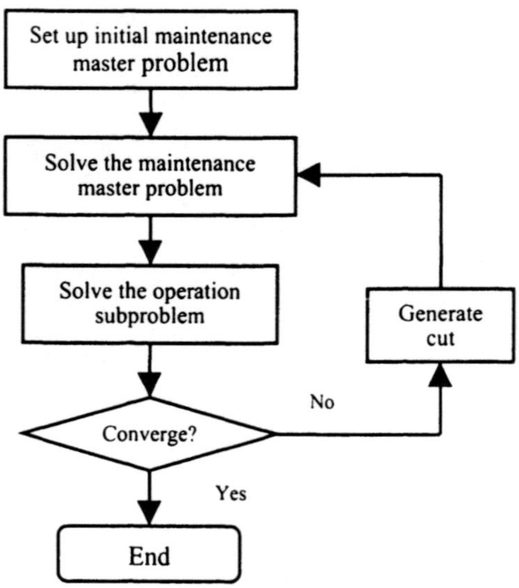

Figure 7.2 Decomposition Method for Line Maintenance Scheduling

7.3.1 Long-term Maintenance Scheduling (TRANSCO).

This problem was discussed thoroughly in Chapter 6. We use the output of long-term maintenance scheduling as an input to short-term maintenance scheduling which is implemented by the ISO.

7.3.2 Short-term Maintenance Scheduling (ISO).

This problem is an operation problem which is handled by the ISO upon receiving the generation information from GENCOs. According to (7.16), the initial short-term maintenance master program is formulated as follows:

$$Min\ z$$

$$St.\ z \geq \sum_{\tau} \left\{ \sum_{k \in \omega} C'_{kt} M_k (1 - y_{k\tau}) \right\}$$

$$\begin{aligned} & y_{k\tau} = 1 && \text{for } y_{kt} = 1 \quad \forall \tau \in t \\ & (Y^{off}_{k(\tau-1)} - h_k)(y_{k\tau} - y_{k(\tau-1)}) \geq 0 && \text{for } y_{kt} = 0 \quad \forall \tau \in t \\ & \sum_{\tau} y_{kt} = n_\tau - h_k && \text{for } y_{kt} = 0 \quad \forall \tau \in t \end{aligned} \tag{7.17}$$

a) Operation Subproblems

The transmission providers' revenues depend on the availability of transmission lines for wheeling power transactions. Thus, the loss of revenue due to maintenance, w_τ, ought to be minimized. Here the ISO will try to minimize the corresponding violations in real and reactive power flow flows. The less violation in flows is translated into more revenues for a TRANSCO as it will be able to meet the demand. The formulation is expressed as follows:

$$Min\ \ w_\tau = \sum_{i \in \Omega} \{c_{i\tau}\ r_{i\tau}\} + \sum_{i \in \beta} \{c'_{i\tau}\ s_{i\tau}\}$$

$$\begin{aligned} St.\ & \boldsymbol{B\theta} + \boldsymbol{r} = \boldsymbol{P} \\ & \boldsymbol{r} \leq \overline{\boldsymbol{r}} \\ & \underline{\boldsymbol{\Psi}} \leq |\boldsymbol{S'\theta}| \leq \overline{\boldsymbol{\Psi}} \\ & \boldsymbol{J''_1}\ \Delta \boldsymbol{V} - \boldsymbol{s} \leq \overline{\boldsymbol{Q}} \\ & \boldsymbol{J''_2}\ \Delta \boldsymbol{V} = \boldsymbol{0} \\ & \Delta \underline{\boldsymbol{V}} \leq \Delta \boldsymbol{V} \leq \Delta \overline{\boldsymbol{V}} \\ & \boldsymbol{s} \leq \overline{\boldsymbol{s}} \end{aligned} \tag{7.18}$$

Network constraints are independent of each other at each time period, and the structure of network constraints makes it possible to solve them independently at each time period. In addition, (7.18) consists of two independent sets of real and reactive flows which could be decomposed into two subproblems as follows:

- Transmission Limit Subproblem

The maintenance decision $\boldsymbol{N}$ in the active power flow model also corresponds to transmission line capacity. However, a transmission line in this model is characterized by two parameters: flow capacity $\bar{f}$ and susceptance γ. Observing (7.18), we notice that γ is not in the right-hand side of the problem, but is part of the $\boldsymbol{B}$ matrix. The susceptance γ can be written explicitly as a decision variable if we rewrite the transmission limit subproblem in (7.18) as given in (7.19).

$$\begin{aligned} Min \quad & wl_{\tau} = \sum_{i \in \Omega} \{c_{i\tau} \, r_{i\tau}\} \\ St. \quad & \boldsymbol{S}\, \boldsymbol{f} + \boldsymbol{r} = \boldsymbol{P} && \text{(dual variable is } \upsilon) \\ & \boldsymbol{r} \le \bar{\boldsymbol{r}} \\ & \underline{\boldsymbol{f}} \le \boldsymbol{f} \le \bar{\boldsymbol{f}} && \text{(dual variable is } \lambda) \\ & \boldsymbol{f} / \boldsymbol{\Psi} = \boldsymbol{\gamma} && \text{(dual variable is } \xi) \end{aligned} \tag{7.19}$$

If we define wl_{τ}^{n} as the loss of revenue at hour τ associated with the nth trial solution, the multiplier λ^{n} may be interpreted as the marginal loss of revenue associated with 1 MW decrease in the line capacity, given the nth trial maintenance schedule. The multiplier ξ^{n} may be interpreted as the marginal loss of revenue associated with a unit decrease in the line susceptance, given the nth trial maintenance schedule. To construct Benders cuts for the master problem (7.17), we have to find multipliers $\lambda^{n}, \xi^{n}, \mu^{n}$. Since (7.18) is non-linear due to $\boldsymbol{f}/\psi$, we solve the following subproblem:

$$\begin{aligned} Min \quad & wl_{\tau} = \sum_{i \in \Omega} \{c_{i\tau} \, r_{i\tau}\} \\ St. \quad & \boldsymbol{B\theta} + \boldsymbol{r} = \boldsymbol{P} && \text{(dual variable } \upsilon) \\ & \boldsymbol{B\theta} \le \bar{\boldsymbol{f}} && \text{(dual variable } \lambda) \\ & \boldsymbol{B\theta} \ge \underline{\boldsymbol{f}} \\ & \boldsymbol{r} \le \bar{\boldsymbol{r}} \\ & \underline{\boldsymbol{\Psi}} \le |\boldsymbol{S}'\boldsymbol{\theta}| \le \bar{\boldsymbol{\Psi}} \end{aligned} \tag{7.20}$$

The multiplier ξ_{k}^{n} that associates with the sensitivity of the susceptance of branch k (between buses i and j) may be calculated as:

$$\xi_{k}^{n}(i,j) = (\upsilon_{i}^{n} - \upsilon_{j}^{n})\,(\theta_{j}^{n} - \theta_{i}^{n}) \tag{7.21}$$

where, θ_i^n and θ_j^n are voltage angles at the optimal solution in iteration n

υ_i^n and υ_j^n are multipliers of the first constraint in (7.20)

- Voltage Limit Subproblem

The maintenance decision $\boldsymbol{y}$ in the reactive power flow model corresponds to $\boldsymbol{J''}$ matrix. Observing (7.18), we notice that $\boldsymbol{J''}$ is not on the right-hand side of the problem. The problem of reactive power is almost similar to that of real power in which $\boldsymbol{J''}$ and $\Delta\boldsymbol{V}$ are used instead of $\boldsymbol{B}$ and θ. First, we solve the reactive power part of the linear problem (7.18) as given in (7.22).

$$
\begin{aligned}
Min \quad & w2_\tau = \sum_{i \in \beta} \{c'_{i\tau}\, s_{i\tau}\} \\
St. \quad & \boldsymbol{J_1''}\, \Delta \boldsymbol{V} - \boldsymbol{s} \le \overline{\boldsymbol{Q}} \qquad \qquad (\textit{dual variable is } \sigma) \\
& \boldsymbol{J_2''}\, \Delta \boldsymbol{V} = \boldsymbol{0} \\
& \Delta \underline{\boldsymbol{V}} \le \Delta \boldsymbol{V} \le \Delta \overline{\boldsymbol{V}} \\
& \boldsymbol{s} \le \overline{\boldsymbol{s}}
\end{aligned}
\tag{7.22}
$$

Instead of finding sensitivity of γ_k which is element (i,j) of $\boldsymbol{B}$, we find sensitivity of J_k'' which is the element (i,j) of matrix $\boldsymbol{J_1''}$. The sensitivity of J_k'', called multiplier μ_k^n, can be calculated using the dual price, σ, of (7.22) as follows:

$$
\mu_k^n(i,j) = (\sigma_i^n - \sigma_j^n)(V_j^n - V_i^n) \tag{7.23}
$$

where V_i^n and V_j^n are voltage magnitudes at the optimal solution in iteration n

σ_i^n and σ_j^n are the multipliers of the first constraints in (7.22)

Given the dual multipliers in (7.21) and (7.23), the associated Benders cut of (7.18) is as follows:

$$
\begin{aligned}
z \ge & \sum_\tau \sum_k \{C'_{k\tau} M_k (1 - y_{k\tau}) - c_{k\tau} f_{k\tau} (\overline{N}_{k\tau} - M_k (1 - y_{k\tau}))\} + \sum_\tau w1_\tau^n + \sum_\tau w2_\tau^n \\
& + \sum_\tau \sum_k \{\lambda_{k\tau}^n \bar{f}_k + \xi_{k\tau}^n \gamma_k + \mu_{k\tau}^n J_k''\}(N_{k\tau}^n - N_{k\tau})
\end{aligned}
\tag{7.24}
$$

where N_{kt}^n is the trial solution of the previous master problem in iteration n and λ^n, ξ^n, μ^n are multipliers in the nth iteration. The cut (7.24) will tend to increase lower bounds of the successive maintenance subproblem solutions.

b) Revised Master Problem

Operation subproblems yield a set of dual multipliers from which a cut is constructed in each iteration. This cut is added to the master-problem to revise the previous master-problem. The master-problem at iteration n is:

$$Min\ z$$

$$St.\ \ z \geq \sum_{t} \left\{ \sum_{k \in \omega} C'_{kt} M_k (1 - y_{k\tau}) \right\}$$

$$y_{k\tau} = 1 \qquad\qquad for\ y_{kt} = 1 \quad \forall \tau \in t$$

$$(Y^{off}_{k(\tau-1)} - h_k)(y_{k\tau} - y_{k(\tau-1)}) \geq 0 \qquad\qquad for\ y_{kt} = 0 \quad \forall \tau \in t$$

$$\sum_{\tau} y_{kt} = n_\tau - h_k \qquad\qquad for\ y_{kt} = 0 \quad \forall \tau \in t$$

$$\sum_{k \in A} \sigma_{mk} M_k (1 - y_{k\tau}) \leq \psi_{m\tau}$$

$$z \geq \sum_{\tau}\sum_{k} \left\{ C'_{k\tau} M_k (1 - y_{k\tau}) - c_{k\tau} f_{k\tau} (\bar{N}_{k\tau} - M_k (1 - y_{k\tau})) \right\} + \sum_{\tau} w1^n_\tau + \sum_{\tau} w2^n_\tau$$
$$+ \sum_{\tau}\sum_{k} \left\{ \lambda^n_{k\tau} \bar{f}_k + \xi^n_{k\tau} \gamma_k + \mu^n_{k\tau} J''_k \right\} (N^n_{k\tau} - N_{k\tau})$$

$$y_{k\tau} \in \{0, 1\} \tag{7.25}$$

7.4 CASE STUDIES

7.4.1 Example 1. We apply the proposed method to the IEEE-RTS given in Chapter 6. The transmission maintenance windows in the long-term horizon are given in Table 7.1, which are the same as that of Case 3 in Table 6.10. In this table, the 12-week horizon is allocated between weeks 18-29. Most of the lines are scheduled for maintenance during the lowest peak load in the 12-week horizon. (see Table A.1).

Table 7.1 Transmission Maintenance Windows (12-week horizon)

Between buses	Week
2-4	21
5-10	22
4-9	27
3-9	26

Next, we proceed to schedule the short-term maintenance. The task of short-term maintenance scheduling is to determine the optimal maintenance intervals within the maintenance week. If the short-term maintenance duration can be divided into shorter (hourly) intervals when the demand is less, the solution of this problem will determine those hours within the given the maintenance week.

In our example, we consider weeks 18-29 as a case of short-term maintenance scheduling. Only lines 2-4, 5-10, 4-9, and 3-9 are scheduled to be on long-term scheduling maintenance. The weekly peak transaction is given in Table 7.2. Load data can be obtained using the percentage of system load in Table A.9. We assume that the maintenance duration for each line is only one hour and the maintenance cost is given in Table A.13. A step-size for reactive problem (see Section 2.6) is used as $\Delta V_{step} = 0.25$, $\Delta Q_{step} = 0.1$. The bus generation schedule was given in Table 3.17. The hourly transaction profile calculated from Tables A.1, A.2 and A.3 of Appendix A is given in Figure 7.3.

Table 7.2 Transaction Weekly Peak

Week	*P* (MW)	*Q* (MVA)
21	2440	488
22	2311	462
27	1947	389
26	2326	465

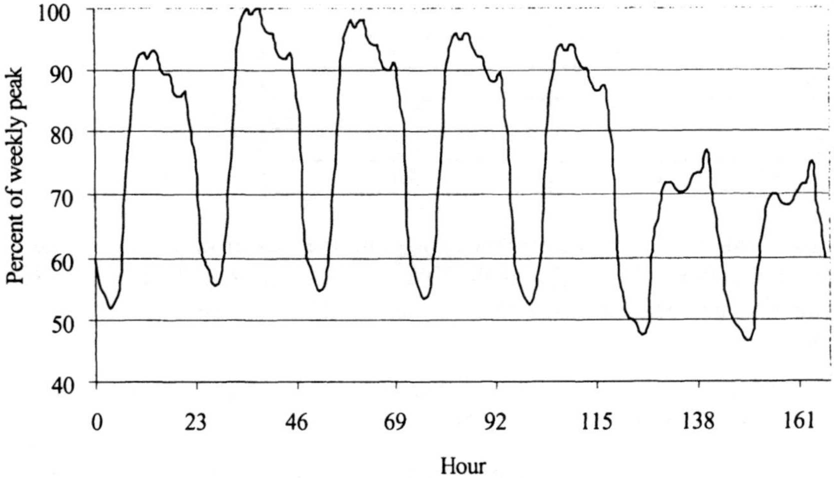

Figure 7.3 Hourly Peak Transaction in Percent of Weekly Peak

The worst transmission contingency is chosen as the outage of line 12-23. Here, the procedure screens all possible transmission contingencies in the system. The line flow violations are accumulated for individual contingencies. The contingency corresponding to the largest violation is defined as the worst contingency case in the system. Transmission providers may relieve their worst contingency by interrupting recallable contracts with a perceived loss of revenue given in Table 7.3.

Table 7.3 Recallable Contracts in Week 18-29

Bus	$\bar{r}+j\,\bar{s}$(%)*	$c_{i\tau}$ (\$/MWh)	$c'_{i\tau}$ (\$/MVAR)
3	2.12	11.87	1.18
6	2.26	8.40	0.84
7	2.05	16.21	1.62
8	2.66	18.47	1.84
9	6.53	8.98	0.90
10	1.6	27.88	2.79
13	3.06	21.40	2.14
15	3.84	17.96	1.80
18	1.60	19.28	1.93
20	4.34	9.53	0.95

* percent of total hourly load

A transmission provider may consider two options to provide the deficient reactive power due to line maintenance. The first option is to cancel the reactive power delivery with a perceived loss of revenue given in Table 7.3. The second option is to purchase reactive power from ancillary service providers. The cost of purchasing additional reactive power is given in Table 7.4

Table 7.4 Ancillary Services for Reactive Power in Week 18-29

Bus	$\bar{s}$(MVAR)	$c'_{i\tau}$ (\$/MVAR)
7	60	1.2
13	80	0.9
18	60	1.0
21	60	0.9
22	32	1.3
23	100	0.8

The minimization of maintenance costs and the loss of revenue are used in the objective function. The results of the following test cases will show the effect of transmission and voltage constraints on short-term transmission line maintenance scheduling.

- Case 1: We consider line maintenance constraints with no transmission flow constraints (TRANSCO).
- Case 2: We consider line maintenance and transmission flow constraints but no voltage constraints (ISO).
- Case 3: In addition to Case 2, voltage constrains are imposed on (ISO).

Case 1 is the classical integer optimization since there are no line flow or voltage constraints. The schedule that gives the minimum maintenance cost is between hours between 9 a.m. and 5 p.m. which, according to Table 7.5, represents the cheapest maintenance cost.

Table 7.5 Hourly Line Maintenance Cost in Weeks 18-29

τ	$C'_{k\tau}$ (10^4\$/km)	τ	$C'_{k\tau}$ (10^4\$/km)	τ	$C'_{k\tau}$ (10^4\$/km)
1	1.00	9	0.72	17	0.72
2	1.00	10	0.72	18	0.80
3	1.00	11	0.72	19	0.80
4	1.00	12	0.72	20	0.80
5	1.00	13	0.72	21	0.80
6	1.00	14	0.72	22	0.80
7	0.80	15	0.72	23	1.00
8	0.80	16	0.72	24	1.00

Since we do not consider line flow and voltage constraints in this case, let's assume that we decide to take line 2-4 out on Monday at 10 a.m. during the week 21. The resulting transmission over-flow is shown in Table 7.6. In this case, there will be flow violations on lines 1-2, 1-3 and 1-5. There are also voltage violations at buses 13, 15 and 18 as seen in Table 7.7. Most of the voltage violations occur on heavily loaded buses.

Table 7.6 Overflow on Transmission Lines (Case 1)

Line Between Buses	MW Rating	Peak MW Flow
1-2	208	211.56
1-3	193	198.12
1-5	208	221.98

Table 7.7 Violated Bus Voltages (Case 1)

Bus No.	Voltage (p.u.)
15	0.853
18	0.829
13	0.895

The imposed transmission constraints increase the cost of maintenance in Case 2. To satisfy transmission constraints, 48 MW of recallable contracts at bus 6 has been interrupted at hour 151 of week 21. The maintenance schedule, which satisfies transmission security constraints, is given in Table 7.8.

Table 7.8 Transmission Maintenance Windows (Case 2)

Between buses	Hour	Week
2-4	151	21
5-10	152	22
4-9	152	27
3-9	151	26

There is no transmission line on maintenance during hours 9-14 due to heavy transactions at these hours. Lines 2-4, 5-10, 4-9 and 3-9 are forced to be on an expensive maintenance schedule to avoid transmission flow violations. The new transmission flows reduce voltage violations but do not guarantee that all bus voltages will be within limits. The new bus voltages are given in Table 7.9. The reason voltages in the local area are too low is that the local area does not have enough reactive power sources. In addition to 48 MW at bus 6 that has been interrupted, 5 MVAR recallable contract at bus 9 has also been interrupted at hour 150 of week 21. Transmission provider also has to purchase 65 MVAR of reactive power at hour 150 from bus 13. The system maintenance cost when voltage limits are not considered is $\$0.9186 \times 10^6$.

Table 7.9 Violated Bus Voltages (Case 2)

Bus No.	Hour	Week	Voltage (p.u.)
15	151	21	0.891
18	151	21	0.899

In Case 3, voltage constraints are considered in maintenance scheduling. As we impose voltage and transmission limits on scheduling, a solution which satisfies both voltage and transmission limits is given in Table 7.10.

Table 7.10 Transmission Maintenance Windows

Between buses	Hour	Week
2-4	150	21
5-10	152	22
4-9	152	27
3-9	150	26

The maintenance cost increases to $\$1.0506 \times 10^6$ as compared to $\$0.9186 \times 10^6$ when voltage limits were not considered as shown in Table 7.11. The maintenance during these hours is more expensive because:

- Maintenance schedule for lines 2-4 and 3-9, which are the two longest lines, is shifted to hour 150.
- Loss of revenue occurs due to canceling some reactive power contracts at bus 9.
- Reactive power is purchased from ancillary service providers for bus 13.

Table 7.11 Line Maintenance Cost

Case	Cost (10^6\$)
1	0.8208
2	0.9186
3	1.0506

Here, we considered one TRANSCO in the case studies. However, in the case of multiple TRANSCOs, the cuts will be divided linearly among entities.

7.4.2 Example 2.

We use the 186-line IEEE-118 bus network (Appendix E) to test the proposed method. Transmission lines have a 300 MW capacity and a step-size for reactive problem is used as $\Delta V_{step} = 0.25$, $\Delta Q_{step} = 0.1$. In our example, the software is executed for a 24-hour period, however, it can easily extended to a week period for practical purposes. Hourly total transactions are given in Table E2. Table E3 gives the data for transmission lines that require maintenance. The short-term maintenance duration for each line is one hour. The maintenance cost for the 10 lines considered in this example is given in Table E4. Transmission providers may relieve a contingency in the system by interrupting recallable contracts with the corresponding loss of revenue given in Table E5. Based on the static security analysis, the worst transmission contingency is chosen as the outage of line 37-38 in this example.

The transmission provider again has two options to provide additional reactive power which may be required due to line maintenance. The first option is to interrupt reactive power delivery. The loss of revenue of this action is given in Table E5. The second option is to purchase reactive power from ancillary services. The cost of purchasing additional reactive power is given in Table E6.

The results of the following test cases are again included to show the effect of transmission and voltage constraints on the line maintenance scheduling.

- Case 1: We consider line maintenance constraints with no network constraints (TRANSCO).
- Case 2: We consider line maintenance and transmission flow constraints but no voltage constraints (ISO).
- Case 3: In addition to Case 2, voltage constrains are imposed on line maintenance.

Similar to Example 1 in Section 7.4.1, Case 1 is a classical integer programming optimization since there are no transmission and voltage constraints. In Case 1, no contract has been cancelled. The final maintenance cost is given in Table 7.12. The corresponding hourly line maintenance schedule is shown in Table 7.13. The numbers (1 or 0) in Table 7.13 represent on/off states of lines for maintenance at different hours: off indicates that the line is unavailable and scheduled for maintenance. It is obvious that the schedule gives the minimum maintenance cost since most lines are scheduled between hours 9 and 17; Table E4 indicates that during these hours the maintenance cost is the cheapest. Now we check the operation constraints.

Table 7.12 Line Maintenance Cost

Case	Cost ($)
0	2292.84
1	4655.64
2	8244.08

Table 7.13 Line Maintenance Schedule without Network Constraints (Case 1)

Line	Hour (1-24)																						
	1					6						12						18						24
1	1	1	1	1	1	1	1	1	1	1	1	0	1	1	1	1	1	1	1	1	1	1	1	1
2	1	1	1	1	1	1	1	1	1	1	1	1	1	1	1	0	1	1	1	1	1	1	1	1
3	1	1	1	1	1	1	1	1	1	1	1	1	1	1	1	1	0	1	1	1	1	1	1	1
4	1	1	1	1	1	1	1	1	1	0	1	1	1	1	1	1	1	1	1	1	1	1	1	1
5	1	1	1	1	1	1	1	1	1	1	0	1	1	1	1	1	1	1	1	1	1	1	1	1
6	1	1	1	1	1	1	1	1	0	1	1	1	1	1	1	1	1	1	1	1	1	1	1	1
7	1	1	1	1	1	1	1	1	1	1	1	1	0	1	1	1	1	1	1	1	1	1	1	1
8	1	1	1	1	1	1	1	1	1	1	1	1	1	0	1	1	1	1	1	1	1	1	1	1
9	1	1	1	1	1	1	1	1	1	1	1	1	1	1	1	1	1	0	1	1	1	1	1	1
10	1	1	1	1	1	1	1	1	1	1	1	1	1	1	0	1	1	1	1	1	1	1	1	1

The maximum transmission flow over the 24-hour study period is shown in Table 7.14. In this case, there are flow violations on lines 8-30, 69-3 and 89-92. There are also voltage violations at buses 107 and 112 as seen in Table 7.15; the violated voltages are shown as bold. Most of these violations occur during heavy transaction periods.

Table 7.14 Overflow on Transmission Lines (Case 1)

Line Between Buses	MW Rating	Peak MW Flow
8-30	300	302.56
69-3	300	300.12
89-92	300	421.98

Table 7.15 Bus Voltages without Constraints in Local Area (Case 1)

Bus No	Voltage (p.u.)					
	Hour 1	Hour 2	Hour 14	Hour 15	Hour 16	Hour 24
104	0.9143	0.9203	0.9177	0.9187	0.9187	0.9475
105	0.9096	0.9165	0.9156	0.9167	0.9167	0.9397
106	0.9014	0.9092	0.9048	0.9060	0.9060	0.9426
107	**0.8854**	**0.8955**	**0.8863**	**0.8879**	**0.8879**	0.9215
108	0.9091	0.9158	0.9336	0.9346	0.9346	0.9295
109	0.9096	0.9161	0.9412	0.9422	0.9422	0.9260
110	0.9159	0.9212	0.9658	0.9666	0.9666	0.9216
112	**0.8882**	**0.8961**	0.9361	0.9373	0.9373	**0.8929**

We now solve Case 2 in our study. The imposed transmission constraints increase the cost of maintenance in Case 2. To satisfy transmission constraints, 110.6 MW of recallable contracts at bus 42 have been interrupted at hours 19 and 21. Table 7.12 shows a change in operating cost over the study period, indicating a shift in the maintenance schedule.

The maintenance schedule which satisfies transmission security constraints, is given in Table 7.16. There are no transmission lines on maintenance during hours 9-14 due to heavy transactions in these hours. Lines 2, 3, 7, 8, 9 and 10 are forced to be on maintenance with additional costs to avoid flow violations in the network (see Table E4). Transmission constraints reduce voltage violations but do not guarantee that all bus voltages will be within limits. The bus voltages after transmission constraints are imposed can be seen in Table 7.17. Again, voltages in a local area are low, as the local area does not have enough reactive power generation.

Table 7.16 Line Maintenance Schedule with Transmission Constraints (Case 2)

Line	Hour (1-24)																						
	1					6						12						18						24
1	1	1	1	1	1	1	1	1	0	1	1	1	1	1	1	1	1	1	1	1	1	1	1	1
2	1	1	1	1	1	1	1	1	1	1	1	1	1	1	1	1	1	1	0	1	1	1	1	1
3	1	1	1	1	1	1	1	1	1	1	1	1	1	1	1	1	1	1	1	1	0	1	1	1
4	1	1	1	1	1	1	1	1	1	1	1	1	1	1	1	0	1	1	1	1	1	1	1	1
5	1	1	1	1	1	1	1	1	1	1	1	1	1	1	0	1	1	1	1	1	1	1	1	1
6	1	1	1	1	1	1	1	1	1	1	1	1	1	1	1	1	0	1	1	1	1	1	1	1
7	1	1	1	1	1	1	0	1	1	1	1	1	1	1	1	1	1	1	1	1	1	1	1	1
8	1	1	1	1	1	1	1	0	1	1	1	1	1	1	1	1	1	1	1	1	1	1	1	1
9	1	1	1	1	1	1	1	1	1	1	1	1	1	1	1	1	1	1	1	1	1	0	1	1
10	1	1	1	1	1	1	1	1	1	1	1	1	1	1	1	1	1	0	1	1	1	1	1	1

In addition of 110.6 MW at bus 42 which have been interrupted, 205 MVAR of recallable contracts at bus 59 have been interrupted at hours 14 and 15. The transmission provider has to purchase reactive power of 138 MVAR at hour 1 and 187 MVAR at hours 14, 15 and 16 from bus 89. System maintenance cost when voltage limits are not considered (Case 2) is $4655.64.

Table 7.17 Bus Voltages without Voltage Constraints in Local Area (Case 2)

Bus No	Voltage (p.u.)					
	Hour 1	Hour 2	Hour 14	Hour 15	Hour 16	Hour 24
104	0.9152	0.9201	0.9172	0.9185	0.9183	0.9572
105	0.9091	0.9161	0.9152	0.9167	0.9164	0.9399
106	0.9015	0.9091	0.9052	0.9066	0.9061	0.9522
107	**0.8953**	0.9211	**0.8952**	**0.8865**	**0.8997**	0.9215
108	0.9091	0.9158	0.9336	0.9356	0.9356	0.9293
109	0.9096	0.9161	0.9512	0.9522	0.9522	0.9261
110	0.9153	0.9211	0.9657	0.9626	0.9631	0.9213
112	**0.8987**	0.9101	0.9366	0.9377	0.9372	0.9029

In Case 3, the effect of voltage constraints on maintenance schedule is considered. As we impose voltage and transmission limits on the scheduling problem, a solution which satisfies both voltage and transmission limits is reached as shown in Table 7.18. The maintenance cost increases to $8244.08 as compared to

$4655.64 when voltages were not considered. The reason that the cost is increased is that the maintenance schedule of lines 4 and 6 which are the two longest lines are shifted to hours 5 and 6 (see Table E4). Table 7.19 shows corrected bus voltages in Case 3.

Table 7.18 Line Maintenance Schedule with Transmission and Voltage Constraints (Case 3)

Line	Hour (1-24)																							
	1					6						12						18						24
1	1	1	1	1	1	1	1	1	0	1	1	1	1	1	1	1	1	1	1	1	1	1	1	1
2	1	1	1	1	1	1	1	1	1	1	1	1	1	1	1	1	1	1	0	1	1	1	1	1
3	1	1	1	1	1	1	1	1	1	1	1	1	1	1	1	1	1	1	1	1	0	1	1	1
4	1	1	1	1	1	0	1	1	1	1	1	1	1	1	1	0	1	1	1	1	1	1	1	1
5	1	1	1	1	1	1	1	1	1	1	1	1	1	1	0	1	1	1	1	1	1	1	1	1
6	1	1	1	1	0	1	1	1	1	1	1	1	1	1	1	1	1	1	1	1	1	1	1	1
7	1	1	1	1	1	1	0	1	1	1	1	1	1	1	1	1	1	1	1	1	1	1	1	1
8	1	1	1	1	1	1	1	0	1	1	1	1	1	1	1	1	1	1	1	1	1	1	1	1
9	1	1	1	1	1	1	1	1	1	1	1	1	1	1	1	1	1	1	1	1	1	0	1	1
10	1	1	1	1	1	1	1	1	1	1	1	1	1	1	1	1	1	0	1	1	1	1	1	1

Table 7.19 Bus Voltages with Voltage Constraints in Local Area (Case 3)

Bus No	Voltage (p.u.)					
	Hour 1	Hour 2	Hour 14	Hour 15	Hour 16	Hour 24
104	0.9682	0.9737	0.9715	0.9723	0.9723	0.9222
105	0.9638	0.9701	0.9696	0.9706	0.9706	0.9211
106	0.9477	0.9548	0.9510	0.9521	0.9521	0.9106
107	0.9379	0.9471	0.9389	0.9404	0.9404	0.9010
108	0.9564	0.9625	0.9800	0.9809	0.9809	0.9375
109	0.9539	0.9598	0.9845	0.9854	0.9854	0.9449
110	0.9517	0.9565	1.0004	1.0011	1.0011	0.9690
112	0.9253	0.9327	0.9721	0.9732	0.9732	0.9421

CHAPTER VIII

COORDINATION BETWEEN GENERATION AND TRANSMISSION MAINTENANCE SCHEDULING

8.1 INTRODUCTION

The basic premise of transmission open access is that transmission owners/providers (TRANSCO) would treat all transmissions users on a non-discriminatory basis regarding the access to and the use of transmission system and services. This requirement could be difficult to ensure if TRANSCOs would have any financial interests in energy generation or supply. A general trend, therefore, has been to designate the ISO to operate the transmission system and facilitate the provision of transmission services. Maintenance of transmission system generally remains the responsibility of TRANSCOs.

The ISO is responsible for the security of the system operation. The responsibility could include the coordination between maintenance scheduling of generation resources and transmission facilities. The ISO's role in maintenance scheduling is limited to ensuring that the submitted schedules do not cause transmission congestion or endanger the system reliability. Under some circumstances, if the requirements are not met, the ISO may have to choose among several scheduling options in a non-discriminatory manner. In this case, the ISO could select schedules that are most economical.

In this chapter, we discuss the maintenance scheduling coordination in restructured power system involving ISO, GENCO and TRANSCO. The proposed method also takes into account two possible network models, e.g. the transportation model and the linearized power flow model.

8.2 PROBLEM FORMULATION

The overall objective is to operate the power system economically and reliably. The schedule submitted to the ISO by GENCOs and TRANSCOs should satisfy security and reliability constraints. Here we are assuming that the ISO is dealing with one GENCO and one TRANSCO. However, in the case of multiple entities, Benders cuts will be divided linearly among those entities. The ISO's feasibility check is formulated as follows:

$$Min\ \ E\left\{\sum_i r_{it}\right\}$$

$$
\begin{aligned}
St.\quad & \boldsymbol{Sf + g + r = d(\varphi)} \\
& \boldsymbol{g \le \bar{g}(\varphi).x} \\
& \boldsymbol{r \le d(\varphi)} \\
& \boldsymbol{|f| \le \bar{f}(\varphi).N}
\end{aligned}
\tag{8.1}
$$

Variable $\boldsymbol{x}$ and $\boldsymbol{N}$ are maintenance schedules that are submitted by GENCOs and TRANSCOs. If the $E\{r_{it}\}$ is larger than a prescribed value, let say ε, then the ISO will inform GENCOs and TRANSCOs to recalculate their schedules. Figure 8.1 depicts the interactions between the ISO, GENCO and TRANSCO.

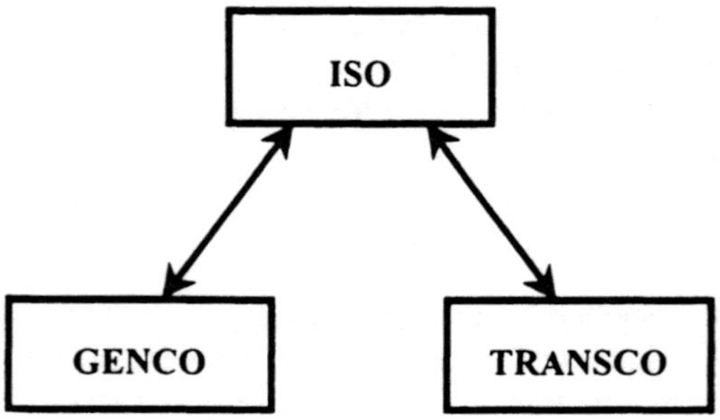

Figure 8.1 Interactions between Entities

Mathematically, if $E\{r_{it}\} > \varepsilon$, the Benders cut will be in the following form:

$$
E\left\{\sum_i r_{it}^n\right\} + \sum_i \lambda_{it}^n \, \bar{g}_i \, (x_{it}^n - x_{it}) + \sum_k \lambda_{kt}^{\prime n} \, \bar{f}_k \, (N_{kt}^n - N_{kt}) \le \varepsilon \tag{8.2}
$$

The problem with constraint (8.2) is that GENCOs do not have any information about the transmission line maintenance schedule and TRANSCOs do not have any information about the generation maintenance schedule.

If reliability requirements are met, the ISO will use an economical criterion to select a schedule, among proposed schedules, that will minimizes the total operation cost.

The formulation for a feasible maintenance schedule is as follows:

$$
w_t = Min \;\; E\left\{\sum_i c_{it} g_{it} + \sum_k c_{kt}^{'} f_{kt}\right\}
$$

$$
\begin{aligned}
St.\quad & Sf + g + r = d(\varphi) \\
& g \le \bar{g}(\varphi).x \\
& r \le d(\varphi) \\
& |f| \le \bar{f}(\varphi).N \\
& \sum_i r_{it} \le \varepsilon(\varphi)
\end{aligned}
\qquad (8.3)
$$

In which the total cost is $w = \sum_t w_t$. The first term in the objective of (8.3) is the total generation cost and the second term is the total transmission charge at time t. The reliability constraint is represented as the fifth constraint.

8.3 SOLUTION METHODOLOGY

Let us consider the following scenario in Figure 8.2 in which there is one GENCO and one TRANSCO. If the TRANSCO submits a schedule to the ISO to take out L1 for preventive maintenance, the GENCO cannot submit a schedule to take out g2 since it will not satisfy the reliability constraint. This circumstance results in a schedule driven by the TRANSCO. On the other hand, if the GENCO schedules g2 to be out for maintenance, the TRANSCO cannot take out L1.

To avoid unfair results, the ISO can examine the two alternatives and come up with the most economical schedule from the operation point of view. Here, we would consider two alternative paths. In path 1, maintenance schedules are driven by the GENCO. In path 2, maintenance schedules are driven by the TRANSCO. The final result would be found by comparing operation costs of paths 1 and 2. These two paths are discussed further as follows.

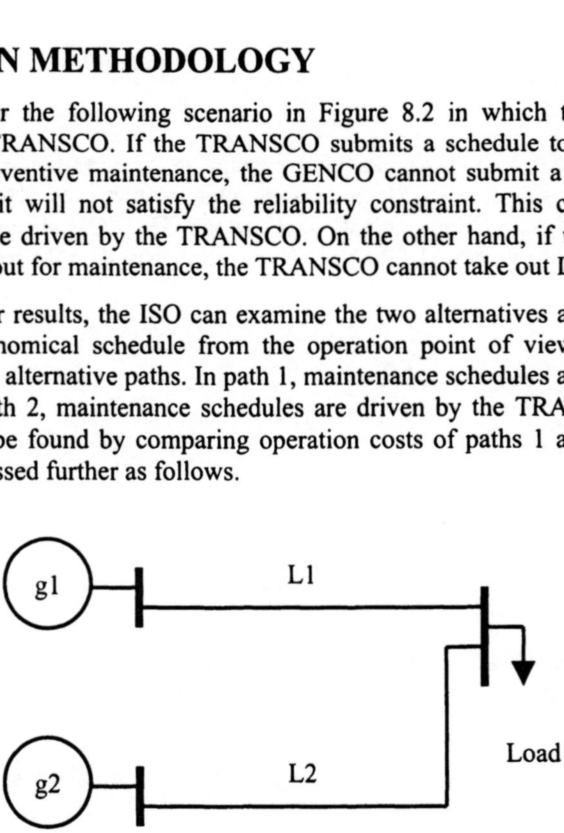

Figure 8.2 Two Scenarios Problem

8.3.1 Path 1: Schedule Driven by the GENCO.

In path 1, the ISO received a generating unit maintenance schedule from the GENCO and calculates the solution for the following problem:

$$Min \;\; E\left\{\sum_i r_{it}\right\}$$

$$
\begin{aligned}
St. \quad & \boldsymbol{Sf + g + r = D(\varphi)} \\
& \boldsymbol{g \le \bar{g}(\varphi).x^n} \qquad\qquad \textit{(dual variable is } \lambda \textit{)} \\
& \boldsymbol{r \le D(\varphi)} \\
& \boldsymbol{|f| \le \bar{f}(\varphi)}
\end{aligned}
\tag{8.4}
$$

If the generating unit maintenance schedule does not satisfy the reliability requirement, then the ISO will send the following Benders cut (constraint) to the GENCO:

$$E\left\{\sum_i r_{it}\right\} + \sum_i \lambda_{it} \, \bar{g}_i \, (x_{it}^n - x_{it}) \le \varepsilon \qquad \forall\, t \in infeasible \tag{8.5}$$

The GENCO needs to add this constraint in its formulation and re-calculate its maintenance schedule. If the generating unit maintenance schedule given by the GENCO is feasible, then the ISO will use the GENCO's schedule to calculate the feasibility of the proposed TRANSCO's schedule based on the following problem:

$$Min \;\; E\left\{\sum_i r_{it}\right\}$$

$$
\begin{aligned}
St. \quad & \boldsymbol{Sf + g + r = D(\varphi)} \\
& \boldsymbol{g \le \bar{g}(\varphi).x^{fix}} \\
& \boldsymbol{r \le D(\varphi)} \\
& \boldsymbol{|f| \le \bar{f}(\varphi).N^n} \qquad\qquad \textit{(dual variable is } \lambda' \textit{)}
\end{aligned}
\tag{8.6}
$$

If the TRANSCO's line maintenance schedule does not satisfy the reliability requirement, then the ISO will send the following Benders cut to the TRANSCO:

$$E\left\{\sum_i r_{it}^n\right\} + \sum_i \lambda_{it}^{'n} \, \bar{f}_k \, (N_{kt}^n - N_{kt}) \le \varepsilon \qquad \forall\, t \in infeasible \tag{8.7}$$

The TRANSCO will add this constraint to its formulation and re-calculate its schedule. If the line maintenance schedule given by TRANSCO is feasible then the operation cost will be calculated as:

$$w_t^1 = Min \;\; E\left\{\sum_i c_{it} g_{it} + \sum_k c_{kt}' f_{kt}\right\}$$

$$
\begin{aligned}
St.\quad & \boldsymbol{Sf+g+r=d(\varphi)} \\
& \boldsymbol{g \le \bar{g}(\varphi).x_l^{fix}} \\
& \boldsymbol{r \le d(\varphi)} \\
& \boldsymbol{|f| \le \bar{f}(\varphi).N_l^{fix}} \\
& \sum_i r_{it} \le \varepsilon(\varphi)
\end{aligned}
\tag{8.8}
$$

8.3.2 Path 2: Schedules Driven by the TRANSCO.

In path 2, the ISO receives the line maintenance schedule from the TRANSCO and calculates the solution for the following problem:

$$
Min\ \ E\left\{\sum_i r_{it}\right\}
$$

$$
\begin{aligned}
St.\quad & \boldsymbol{Sf+g+r=D(\varphi)} \\
& \boldsymbol{g \le \bar{g}(\varphi)} \\
& \boldsymbol{r \le D(\varphi)} \\
& \boldsymbol{|f| \le \bar{f}(\varphi).N^n} \qquad \textit{(dual variable is } \lambda' \textit{)}
\end{aligned}
\tag{8.9}
$$

If the line maintenance schedule does not satisfy the reliability requirement, then the ISO will send the following constraint to the TRANSCO:

$$
E\left\{\sum_i r_{it}^n\right\} + \sum_k \lambda_{kt}'^n \, \bar{f}_k \, (N_{kt}^n - N_{kt}) \le \varepsilon \qquad \forall\, t \in infeasible \tag{8.10}
$$

The TRANSCO will add this constraint to its formulation and re-calculate its schedule. If the line maintenance schedule given by the TRANSCO is feasible, then the ISO will calculate the feasibility of the GENCO's maintenance schedule based on the solution of the following problem:

$$
Min\ \ E\left\{\sum_i r_{it}\right\}
$$

$$
\begin{aligned}
St.\quad & \boldsymbol{Sf+g+r=D(\varphi)} \\
& \boldsymbol{g \le \bar{g}(\varphi).x^n} \qquad \textit{(dual variable is } \lambda \textit{)} \\
& \boldsymbol{r \le D(\varphi)} \\
& \boldsymbol{|f| \le \bar{f}(\varphi).N^{fix}}
\end{aligned}
\tag{8.11}
$$

If the generating unit maintenance schedule does not satisfy the reliability requirement, then the ISO will send the following constraint to the GENCO:

$$E\left\{\sum_i r_{it}\right\} + \sum_i \lambda_{it}\, \bar{g}_i\, (x_{it}^n - x_{it}) \leq \varepsilon \qquad \forall\, t \in infeasible \tag{8.12}$$

The GENCO will add this constraint to its formulation and re-calculate its schedule. If the unit maintenance schedule given by the TRANSCO is feasible, then the operation cost is calculated as:

$$w_t^2 = Min\ E\left\{\sum_i c_{it} g_{it} + \sum_k c'_{kt} f_{kt}\right\}$$

$$\begin{aligned}
St.\quad & \boldsymbol{Sf} + \boldsymbol{g} + \boldsymbol{r} = \boldsymbol{d}(\varphi) \\
& \boldsymbol{g} \leq \bar{\boldsymbol{g}}(\varphi).\boldsymbol{x}_2^{fix} \\
& \boldsymbol{r} \leq \boldsymbol{d}(\varphi) \\
& |\boldsymbol{f}| \leq \bar{\boldsymbol{f}}(\varphi).\boldsymbol{N}_2^{fix} \\
& \sum_i r_{it} \leq \varepsilon(\varphi)
\end{aligned} \tag{8.13}$$

8.3.3 Final Maintenance Schedules.

The total operation cost of path 1 is $w^1 = \sum_t w_t^1$ and total operation cost of path 2 is $w^2 = \sum_t w_t^2$. If $w^1 < w^2$, then schedules $\boldsymbol{x}_1^{fix}$ and N_1^{fix} will be approved by the ISO as final schedules. On the other hand, If $w^1 > w^2$, then schedules $\boldsymbol{x}_2^{fix}$ and N_2^{fix} will be approved as final schedules. In the case of two GENCOs and one TRANSCO submitting their schedule simultaneously, the number of paths is 6 as given in Table 8.1. The number of paths is $n!$ where n is the number of entities who submit schedules simultaneously.

Table 8.1 Paths for 2 GENCOs and a TRANSCO

Steps	Path 1	Path 2	Path 3	Path 4	Path 5	Path 6
1	GENCO1	GENCO1	TRANSCO	TRANSCO	GENCO2	GENCO2
2	GENCO 2	TRANSCO	GENCO1	GENCO2	TRANSCO	GENCO1
3	TRANSCO	GENCO2	GENCO2	GENCO1	GENCO1	TRANSCO

The proposed coordination among GENCO, TRANSCO and ISO is given in Figure 8.3. The number of paths could increases if more than one GENCO and TRANSCO submit their schedules simultaneously.

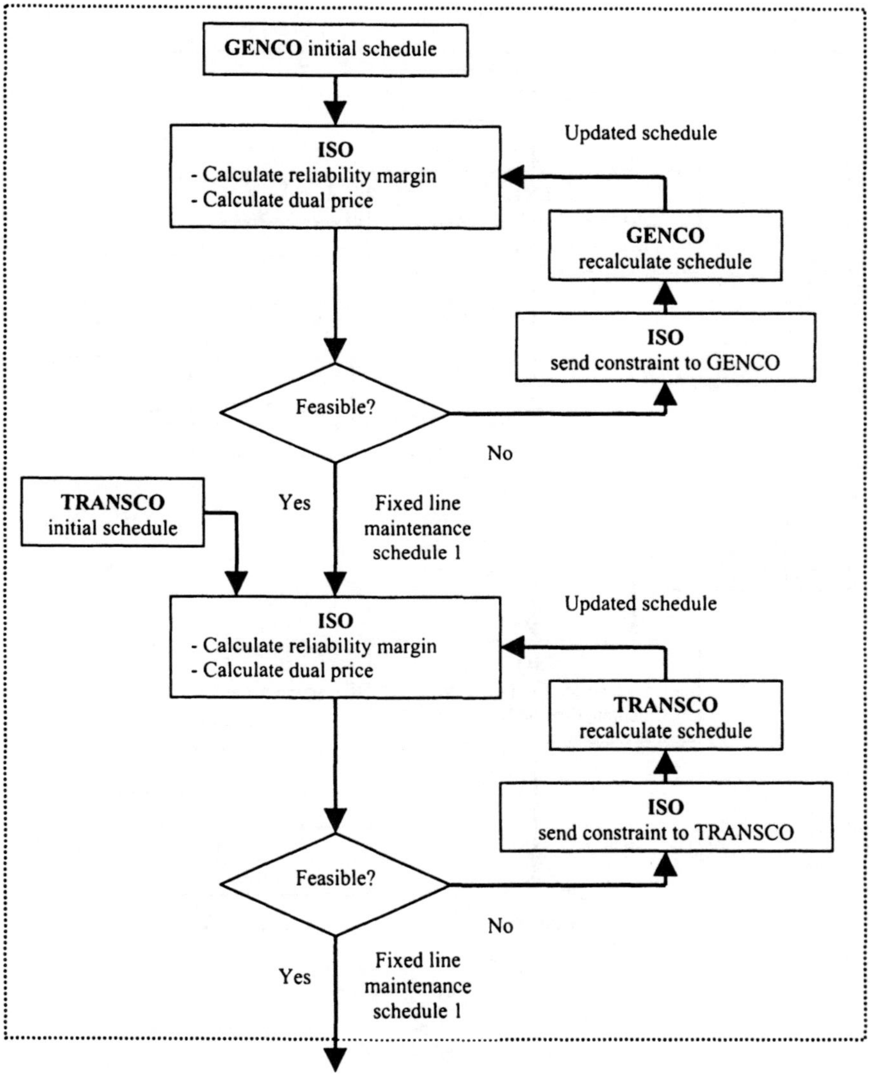

Figure 8.3a Path 1

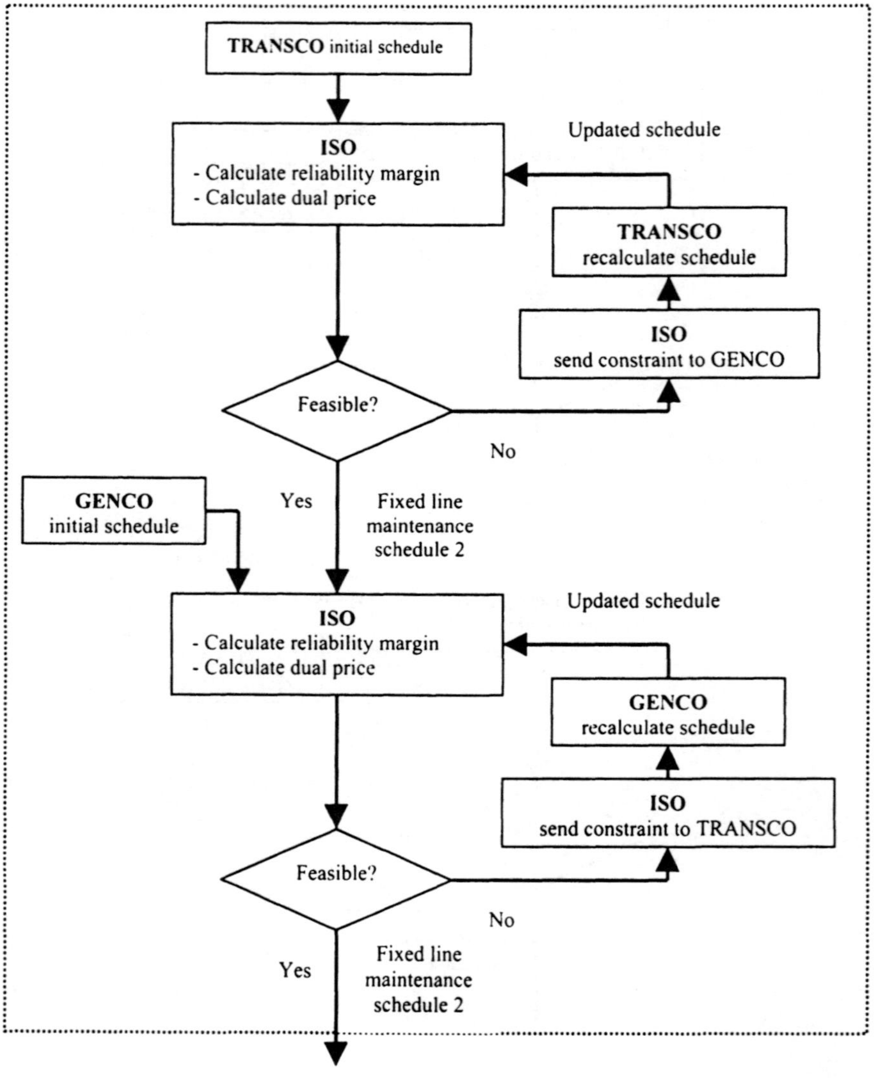

Figure 8.3b Path 2

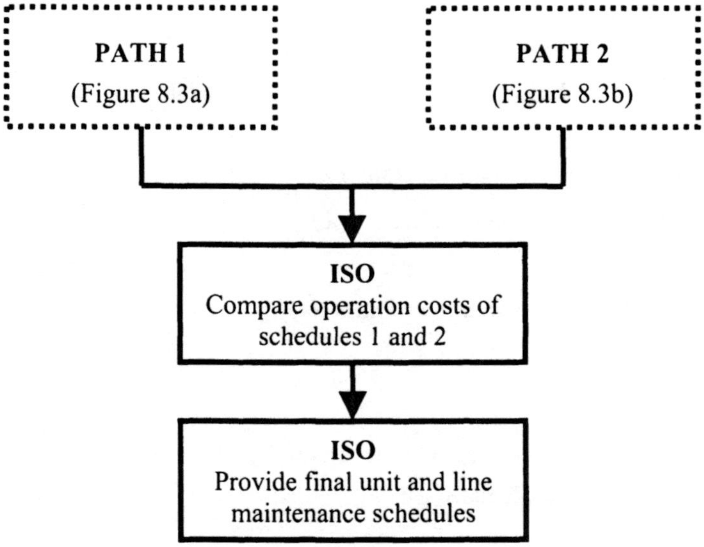

Figure 8.3c Proposed Coordination among GENCO, TRANSCO and ISO

8.3.4 Linearized Power Flow Model. The network can be represented by either a transportation model or a linearized power flow. In the transportation model, given by (8.1), the network obeys only the Kirchoff's first law. The linearized power flow model for (8.1) is as follow:

$$Min \ E\left\{\sum_i r_{it}\right\}$$

St.

$$\begin{aligned} & \boldsymbol{Sf} + \boldsymbol{g} + \boldsymbol{r} = \boldsymbol{d}(\varphi) && \textit{(dual variable is } \upsilon) \\ & \boldsymbol{g} \le \bar{\boldsymbol{g}}(\varphi).\boldsymbol{x} && \textit{(dual variable is } \lambda) \\ & \boldsymbol{r} \le \boldsymbol{d}(\varphi) && \\ & \boldsymbol{f} \le \bar{\boldsymbol{f}}(\varphi).\boldsymbol{N} && \textit{(dual variable is } \lambda') \\ & \boldsymbol{f} / \Psi = \gamma && \textit{(dual variable is } \xi) \end{aligned} \qquad (8.14)$$

Equation (8.14) is similar to the formulation of short-term maintenance scheduling in Chapter 7. Multipliers λ^n, ξ^n, μ^n also can be found using the proposed method in Chapter 7. Since sub-problem (8.14) is non-linear due to $\boldsymbol{f}/\psi$, we solve the following problem:

$$Min\ E\left\{\sum_i r_{it}\right\}$$

St.

$$\begin{aligned} &\boldsymbol{B\theta + g + r = d(\varphi)} && (dual\ variable\ is\ \upsilon) \\ &\boldsymbol{g \le \overline{g}(\varphi).x} && (dual\ variable\ is\ \lambda) \\ &\boldsymbol{r \le d(\varphi)} && \\ &\boldsymbol{B\theta \le \overline{f}(\varphi).N} && (dual\ variable\ is\ \lambda') \\ &\boldsymbol{\underline{\Psi} \le |S'\theta| \le \overline{\Psi}} && \end{aligned} \tag{8.15}$$

The multiplier ξ_k^n that associates with the sensitivity of the susceptance of branch *k* (between buses *i* and *j*) is calculated as:

$$\xi_k^n(i,j) = (\upsilon_i^n - \upsilon_j^n)(\theta_j^n - \theta_i^n) \tag{8.16}$$

where θ_i^n and θ_j^n are voltage angles at the optimal solution in iteration *n*

υ_i^n and υ_j^n are the multipliers of the first set of constraints in (8.15)

Like the transportation model, if the line maintenance schedule does not satisfy the reliability requirement, then the ISO will send the following constraint to the TRANSCO:

$$E\left\{\sum_i r_{it}^n\right\} + \sum_k (\lambda_{kt}^{'n}\,\overline{f}_k + \xi_{kt}^n \gamma_k)(N_{kt}^n - N_{kt}) \le \varepsilon \qquad \forall t \in infeasible \tag{8.17}$$

If the unit maintenance schedule does not satisfy the reliability requirement, then the ISO will send the following constraint to the GENCO:

$$E\left\{\sum_i r_{it}\right\} + \sum_i \lambda_{it}\,\overline{g}_i\,(x_{it}^n - x_{it}) \le \varepsilon \qquad \forall t \in infeasible \tag{8.18}$$

Calculating the following operation cost, the ISO could compare the maintenance schedules for different paths to arrive at the final schedule.

$$w_t = Min\ E\left\{\sum_i c_{it} g_{it} + \sum_k c_{kt}' f_{kt}\right\}$$

St.

$$\begin{aligned}
&B\theta + g + r = d(\varphi) && \textit{(dual variable is } \upsilon\textit{)} \\
&g \le \overline{g}(\varphi).x && \textit{(dual variable is } \lambda\textit{)} \\
&r \le d(\varphi) && \\
&B\theta \le \bar{f}(\varphi).N && \textit{(dual variable is } \lambda'\textit{)} \\
&\underline{\Psi} \le \left| S'\theta \right| \le \overline{\Psi} &&
\end{aligned} \tag{8.19}$$

8.4 EXAMPLE

We consider forced outage rates of generators in the three-bus system given in Figure 8.4. For convenience, in this sample, all lines are assumed to be perfectly reliable. The forced outrage rate (FOR) of each generator is given in Table 8.2. The system reliability requirement (ε) is 0.5 p.u. We assume the study period has only one time interval. Loads are assumed constant during the study period.

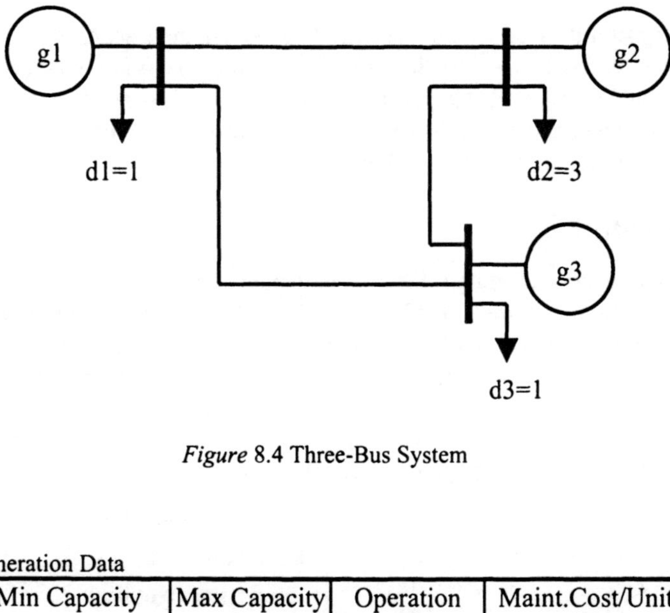

Figure 8.4 Three-Bus System

Table 8.2 Generation Data

Unit	Min Capacity (p.u.)	Max Capacity (p.u.)	Operation Cost ($)	Maint.Cost/Unit ($/Unit)	FOR
1	0.5	2.5	10 g_1	300	0.05
2	0.6	2.5	20 g_2	200	0.10
3	0.6	3.0	30 g_3	100	0.02

Table 8.3 Transmission Line Data

Line	Ω/line	# of lines	Maint.Cost/line ($/line)	Charge ($/p.u)	Max Capacity (p.u.)
1-2	0.20	2	100	0.4	0.25
2-3	0.25	2	200	0.5	0.50
1-3	0.40	2	300	0.6	0.25

Step 1 of Path 1

The GENCO's initial maintenance schedule is: $x_1 = 1, x_2=1, x_3=0$. This schedule is not feasible since $E\{\mathbf{r}\}=0.85$. The ISO will then send a signal to the GENCO which is in the following form:

$$0.85 + 0.4725*2.5*(1-x_2) + 0.56105*3*(0 - x_3) \le 0.5.$$

Adding the above constraint to the GENCO, the GENCO will re-calculate its maintenance schedule as follows: $x_1 = 1, x_2=0, x_3=1$. Given the updated schedule, ISO perform feasibility check. The feasibility check gives $E\{\mathbf{r}\}=1.581$ which is still larger than $\varepsilon=0.5$. The ISO generate the following additional constraint:

$$1.581 + 0.9*2.5*(0-x_2) \le 0.5$$

The new constraint is added to the GENCO which then sends back the updated schedule: $x_1 = 0, x_2=1, x_3=1$ to the SO. Given the trial maintenance schedule in iteration 3, the feasibility check results in $E\{\mathbf{r}\}=0.3$ which is smaller than $\varepsilon=0.5$. This generation maintenance schedule will then be used in Step 2 of path 1 to determine the transmission line maintenance schedule.

Step 2 of Path 1

Given that g_1 is scheduled for a maintenance outage, any line maintenance schedule submitted by the TRANSCO will be rejected by the ISO since the reliability requirement will not be satisfied. The maintenance schedules for path 1 is: $x_1= 0, x_2=1, x_3=1, N_{12}= 2, N_{23}=2, N_{13}=2$. The operation cost for the above schedule is as follow:

$$Min \;\; w= \mathbf{E}\{20*g_2+30*g_3+0.4*f_{12}+0.5*f_{23}+0.6*f_{13}\}$$

St.	$-f_{12}-f_{13}+g_1+r_1=1$	Load balance at bus 1
	$-f_{23}+f_{12}+g_2+r_2=3$	Load balance at bus 2
	$f_{13}+f_{23}+g_3+r_3=1$	Load balance at bus 3
	$0.0 \le g_1 \le 0.0$	Generator 1 limit
	$0.6 \le g_2 \le 2.5$	Generator 2 limit
	$0.6 \le g_3 \le 3.0$	Generator 3 limit
	$-2*0.25 \le f_{12} \le 2*0.25$	Line 1-2 flow limit
	$-2*0.25 \le f_{13} \le 2*0.25$	Line 1-3 flow limit
	$-2*0.5 \le f_{23} \le 2*0.5$	Line 2-3 flow limit

Using the same method given in Chapter 4, the solution of the above problem is given as $119.4604.

Step 1 of Path 2

The TRANSCO's initial maintenance schedule is: $N_{12}=0$, $N_{23}=2$, $N_{13}=2$. Given the trial maintenance schedule by the TRANSCO, the ISO will calculate the amount of reliability. The feasibility check gives $E\{\boldsymbol{r}\}=0.247$ which is smaller than $\varepsilon=0.5$. This line maintenance schedule then will be used in Step 2 of path 2 to determine the line maintenance schedule.

Step 2 of Path 2

Given that line 1-2 is scheduled for a maintenance outage, any unit maintenance schedule submitted by the GENCO will be rejected by the ISO since the reliability requirement will not be satisfied. The maintenance schedules for path 1 is: $x_1=1$, $x_2=1$, $x_3=1$, $N_{12}=0$, $N_{23}=2$, $N_{13}=2$.

The operation cost for the above schedule is as follow:

$$Min\ \ w = \boldsymbol{E}\{10*g_1+20*g2+30*g_3+0.5*f_{23}+0.6*f_{13}\}$$

St.	$-f_{12}-f_{13}+g_1+r_1=1$	Load balance at bus 1
	$-f_{23}+f_{12}+g_2+r_2=3$	Load balance at bus 2
	$f_{13}+f_{23}+g_3+r_3=1$	Load balance at bus 3
	$0.0 \le g_1 \le 2.5$	Generator 1 limit
	$0.6 \le g_2 \le 2.5$	Generator 2 limit
	$0.6 \le g_3 \le 3.0$	Generator 3 limit
	$0.0 \le f_{12} \le 0.0$	Line 1-2 flow limit
	$-2*0.25 \le f_{13} \le 2*0.25$	Line 1-3 flow limit
	$-2*0.5 \le f_{23} \le 2*0.5$	Line 2-3 flow limit

The solution of above problem is $82.1592. Comparing the results of paths 1 and 2, the ISO will find that path 2 is cheaper than path 1. The final maintenance schedule is: $x_1=1$, $x_2=1$, $x_3=1$, $N_{12}=0$, $N_{23}=2$, $N_{13}=2$.

8.5 CASE STUDIES

We apply the same system as the one in Chapters IV and VI in this study. In Section 4.13 of Chapter IV, we considered a case of generation unit maintenance schedule, and in Section 6.4 of Chapter VI we considered a case of transmission maintenance schedule. In this section, we discuss the coordination between those two cases in a restructured power system.

We apply the proposed multi-path method as a tool for the ISO to coordinate maintenance schedules submitted by the GENCO and the TRANSCO. In this case study, we assume that the GENCO and the TRANSCO plan for maintenance in a 3-

month study horizon. Again, we assume that there is only one GENCO and one TRANSCO in the system, and in the case of multiple entities will distribute the Benders cut linearly among the entities.

The results of the following test cases are included to show the effect of coordination on generating units and transmission lines maintenance schedules.

- Case 1: We assume that only the GENCO submits its generating unit maintenance schedule to the ISO.
- Case 2: We assume that only the TRANSCO submits its transmission line maintenance schedule to the ISO.
- Case 3: We consider that both the GENCO and the TRANSCO submit their maintenance schedules to the ISO.

Case 1, is basically the same as Case 4 in Section 4.13 of Chapter IV. In this Case, the GENCO and the ISO coordinate the unit maintenance schedule. The ISO requires the maintenance schedule to satisfy the network reliability requirement. At the same time, the GENCO requires the maintenance schedule to satisfy maintenance constraints, fuel constraints, emission constraints and short-term scheduling constraints in order to achieve a minimum maintenance cost. Table 8.4 gives the unit maintenance schedule. Although some units are not scheduled for maintenance, they may be scheduled to be off at particular weeks because of emission, fuel constraints or economical reasons.

Table 8.4 Generating Unit Maintenance Schedule Case 1 (12-week horizon, ε=1% of load)

Case	Unit	Weeks on Maintenance	Weeks On	Weeks Off
1	1	26,27	19-25,28,29	18
	2	28,29	18-20,22-27	21
	3	18,19	20,21,24-29	22,23
	4	20,21	18,19,22-24,28,29	25-27
	5	27,28	18,21-24	19,20,25,26,29

In Case 2, we apply the proposed method to the system in Chapter VI. The transmission maintenance windows in long-term horizon are given in Table 8.5, which is the same as Case 3 in Section 6.8 of Chapter VI. In this table, the 12-week horizon is between weeks 18-29. Most of the lines are scheduled for maintenance in the lowest peak load of the 12-week horizon. (see Table A.1). In this case, the TRANSCO and the ISO coordinate the transmission line maintenance schedule. The ISO requires the final schedule to satisfy the network reliability requirement. Likewise, the TRANSCO requires the final schedule to satisfy maintenance constraints, short-term scheduling constraints in order to achieve a minimum maintenance cost. Table 8.5 gives the final transmission line maintenance schedule.

Table 8.5 Transmission Maintenance Schedule Case 2 (12-week horizon, ε=1% of load)

Case	Line between buses	Week on Maintenance
2	2-4	21
	5-10	22
	4-9	27
	3-9	26

Case 3 studies the effect of coordination on maintenance schedule. In path 1, the TRANSCO schedules its line maintenance given the fact that the generation unit maintenance schedule has been settled. Given the GENCO's schedule in Table 8.4, the ISO and the TRANSCO coordinate the transmission line maintenance schedule. The transmission line maintenance schedule that has been driven by the GENCO is given in Table 8.6.

Table 8.6 Transmission Maintenance Schedule Driven by GENCO (12-week horizon, ε=1% of load)

Case	Line between buses	Week on Maintenance
3	2-4	21
	5-10	27
	4-9	24
	3-9	26

Comparing Cases 2 and 3, the ISO learns that the maintenance schedule for transmission line 5-10 has been shifted from week 22 to 27. Line 5-10 is forced to be shifted because there is not enough generation to supply the load pocket around bus 10. The load pocket around bus 10 is basically supplied by unit 1 in bus 1 via line 5-10 and units 3 and 4 in bus 7 via lines 7-8 and 8-10. If line 5-10 is on outage at week 22, then unit 1 would not be able to deliver any power to that pocket. Unit 3 that is expected to pick up the generation is also scheduled to be off on week 22. Therefore, the scheduled maintenance of line 5-10 is shifted to week 27 when units 3 and 4 are available to pick up the shortage of unit 1 at that week. Unit 2 via line 4-9 and units 3 and 4 via lines 7-8 and 8-9 supply the load at bus 9. The maintenance schedule for line 4-9 is forced to be shifted from week 27 to 24, since unit 4 that is expected to pick up the shortage of unit 2 is unavailable in week 27.

In path 2, the GENCO schedules its generation unit maintenance given the fact that the line maintenance schedule has been settled. Given the TRANSCO's schedule in Table 8.5, the ISO and the GENCO coordinate the generation unit maintenance schedule. The generation unit maintenance schedule that has been driven by the TRANSCO is seen in Table 8.7.

Table 8.7 Generating Unit Maintenance Schedule Driven by TRANSCO (12-week horizon, ε=1% of load)

Case	Unit	Weeks on Maintenance	Weeks Up	Weeks Down
3	1	26,27	19-25,28,29	18
	2	28,29	18-20,22-27	21
	3	18,19	20,21,22,24-29	23
	4	20,21	18,19,22-24,27-29	25,26
	5	27,28	18,21-24	19,20,25,26,29

Comparing Cases 1 and 3, the ISO learns that the maintenance schedules do not change, only the generation unit availability is changed. Unit 3 that was scheduled to be down on week 22 (see Table 8.4) has been forced to be on line. Unit 3 is required to be available on week 22 since Line 5-10 is scheduled to be on maintenance during the week. Unit 4 has also been forced to be available on week 27 since line 4-9 is scheduled to be on maintenance.

The final maintenance schedule is determined by comparing the operation costs of paths 1 and 2. The operation cost of path 1 is $\$36.69147 \times 10^6$ as compared with $\$37.12461 \times 10^6$ for path 2. Since the operation cost of path 1 is cheaper than that of path 2, the final maintenance schedule is that of path 1 as given in Tables 8.8 and 8.9.

Table 8.8 Final Generating Unit Maintenance Schedule (12-week horizon, ε=1% of load)

Case	Unit	Weeks on Maintenance	Weeks Up	Weeks Down
3	1	26,27	19-25,28,29	18
	2	28,29	18-20,22-27	21
	3	18,19	20,21,24-29	22,23
	4	20,21	18,19,22-24,28,29	25-29
	5	27,28	18,21-24	19,20,25,26,29

Table 8.9 Final Transmission Maintenance Schedule Case 3 (12-week horizon, ε=1% of load)

Case	Line between buses	Week on Maintenance
3	2-4	21
	5-10	27
	4-9	24
	3-9	26

Here, we considered one GENCO and one TRANSCO in the case studies. However, in the case of multiple GENCOs and TRANSCOs, the corresponding cuts will be divided linearly among entities.

CHAPTER IX

APPLICATION OF SHORT-TERM SCHEDULING TO PHOTOVOLTAIC-UTILITY GRID

9.1 INTRODUCTION

Among alternatives that are investigated for the generation of electricity are a number of unconventional sources including solar and wind energy. In many parts of the globe, photovoltaic (PV) systems are considered as a viable alternative for generating electricity. In recent decades, PV has found its first commercial use in space; as hardware costs decline, terrestrial applications of PV would become more economical and perhaps more viable for applications to bulk power generation. Achieving this goal would require GENCOs to evaluate the PV's impact on peak load shaving, expansion planning, and backup generating capacity to ensure a reliable supply of electricity.

As the storage (battery) technology matures and becomes available for the electric utility's load leveling, there may be ways to make batteries more viable by integrating them with PV energy systems. With the availability of advanced batteries, it will be possible to store large amounts of energy during off-peak periods for use during peak hours. However, because of the intermittent nature of the PV generation, a dedicated battery may be considered in the design of integrated systems.

In this chapter, we present a method to simulate the production cost of an integrated photovoltaic-thermal utility with battery storage, based on a probabilistic approach. This methodology would allow us to model solar plants, with different operating characteristics, in a GENCO without a significant modification to the conventional production cost model. Section 9.3 presents a method for estimating PV generation using the statistical description of long-term meteorological data. In Section 9.4, the production cost model of GENCOs is extended to PV generators, using the Equivalent Energy Function (EEF) [Wan88]. In Section 9.5, a numerical example illustrates the usefulness of the proposed method; a pilot project in the Eastern Islands of Indonesia is evaluated using the proposed method.

In the second part of this chapter, we present a method for the short-term scheduling of an integrated PV-thermal system considering the effect of battery storage. The results of this chapter will complement the generation scheduling results given in Chapter 4. To obtain a fast and efficient solution, we utilizes both the successive DP approximation method. The successive DP is used to find the minimum cost trajectory for battery, and the augmented Lagrangian is used to find unit commitment in a GENCO. Maximum PV penetration, battery constraints as well as the availability of natural solar resources are considered in the scheduling formulation.

9.2 PROBABILISTIC PRODUCTION COST FOR PHOTOVOLTAIC-UTILITY SYSTEMS

The prediction of the cost of generation is an important issue which is used for long-term planning, fuel budgeting and system operation. When PV plants represent a part of a GENCO, the production cost simulations are used to model the existing policies for the operation of PV plants. In this regard, long-term planning packages may not be applicable as they model the generation according to a scheme prescribed by control centers. However, because of fluctuations in solar irradiance occurring in the minute-to-minute time frame, PV systems require a special treatment in generation scheduling [Yam81, Cho91, Day82]. In the technical literature, the PV power generation is generally treated as a negative load with the expectation that the electric utility will meet residual load demands with conventional generating sources. This is probably true when the penetration of the PV capacity in the grid is insignificant. With the expected increase in PV-battery's penetration into the grid, the simple treatment of PV as a negative load may not be adequate.

Various analytical and computer models have been suggested for evaluating PV power. The first option is based on a deterministic approach [Ofr83, Kle87, Bar84] which uses a set of parameters for temperature and PV characteristics, among others, to model the power of PV. These models are not very suitable for reliability analysis, production cost simulation and generation expansion planning. The other option is based on a probabilistic approach [Kha86, Abo91] for capturing uncertainties in solar radiation and modeling the performance of PV. However, some of the PV models may not be compatible with conventional production cost simulation packages used in utilities.

9.3 RADIATION AND PV POWER

For operation and control purposes, electric utilities with PV plants are interested in the availability of PV power on a daily basis. The assessment of PV power in electric utilities begins with modeling the solar radiation. The availability and collection of data on insolation, temperature, wind speed and cloud cover at a particular site is vital. Since, the long-term meteorological data are generally available in the form of radiation data on a horizontal surface, the corresponding calculation for a tilted surface will be needed. Various methods have attempted to model the solar irradiance. In this chapter, we compute a tilted surface radiation using the following steps:

Step 1. Calculate the global radiation outside the atmosphere on a horizontal surface (G_{oh}).

Step 2. Calculate the diffuse fraction of the global radiation according to the Orgill/Holands [Org77] using the available long-term radiation data on a horizontal plane (G_h) and G_{oh} from Step 1.

Step 3. Calculate the position of the sun, such as azimuth, elevation, declination and hour angle.

Step 4. Using the output from Steps 2 and 3, calculate the radiation on a tilted surface (G_t) from the available radiation data on a horizontal plane. The solar radiation is time dependent as fluctuations in solar irradiance may occur in a short time.

The detailed calculation is found in Appendix C. Now, let us discuss the PV cell model that is used in conjunction with tilted surface radiation data [Ara82]. The calculation of the output current of a PV generator is dependent on the PV generator voltage, radiation on the PV generator, wind speed and ambient temperature. The relationship between voltage V_c of a cell of PV and current I_c is given by a diode model as:

$$I_C = I_L - I_o\left(exp\left(\frac{q(V_C + I_C R_S)}{AkT}\right) - 1\right) - \frac{(V_C - I_C R_S)}{R_{sh}} \tag{9.1}$$

Equation (9.1) is only applicable to a particular radiation and cell temperature. The light-generated current (I_L) is proportional to the global radiation (G_t) on the generator plane and is assumed to be linearly dependent on the cell temperature T as:

$$I_L = (C_o + C_1 T)\, G_t \tag{9.2}$$

The dependence of the saturation current on temperature is given by:

$$I_o = C_{01} T^3 exp\left(-\frac{1.12\,q}{kT}\right) \tag{9.3}$$

The PV cell temperature is calculated as:

$$T = T_{amb} + K_o - K_1\, v_w \tag{9.4}$$

For a PV generator with N_p parallel cells and N_s cells connected in series, the resulting voltage and current are calculated as:

$$V = N_s V_c \tag{9.5}$$

$$I = N_p I_c \tag{9.6}$$

The PV cell generated power as a function of voltage at 25°C and 1000 W/m2 is given in Figure 9.1. This PV cell has A=1, R_{sh} = 0.130 Ωm^2, R_s=0.381Ωm^2, C_o =0.306V^{-1}, C_1= 0.179x10^{-4} $V^{-1}K^{-1}$, C_{o1}= 1.708x 10^{-4} $Am^{-2}K^{-3}$, Ac=0.01m^2, N_s=36, N_p=1.

The maximum power of the PV cell in Figure 9.1 may be found iteratively; in our case, $\overline{P}(1000)$=38.62W. If we assume that the PV is able to produce 0.9 times its maximum power, then the efficiency of PV as a function of radiation is given as:

$$\eta(\tau) = 0.9\frac{\overline{P}(G_t)}{G_t\, Np\, Ns\, A_c} \tag{9.7}$$

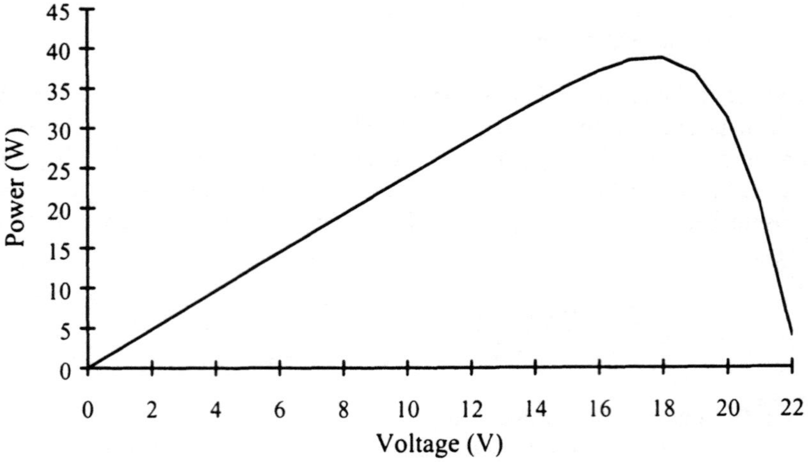

Figure 9.1 Power vs. Voltage Characteristic for a PV Generator

The efficiency of PV for different radiation intensities is depicted in Figure 9.2. At first, a small increase in the radiation produces a significant increase in the PV's efficiency. After a certain radiation point, called K_c, further increases in radiation produce relatively smaller increases in the efficiency. Let us denote the corresponding efficiency as η_c, then the instantaneous PV power can be approximated as:

$$P_{pv}(\tau) = \begin{cases} \frac{\eta_c}{K_c}(G_t)^2 & 0 < G_t < K_c \\ \eta_c G_t & G_t > K_c \end{cases} \tag{9.8}$$

and the daily average power of PV is calculated as:

$$\overline{P}^0(i) = \frac{\int_{a\ day} P_{pv}(\tau)\, d\tau}{\int_{a\ day} d\tau} \tag{9.9}$$

The battery storage plays an important role in the PV system availability. The nominal battery capacity is represented by the long-term and the short-term battery capacity. The former, denoted as storage days, represents the consecutive number of days that a battery is required to supply a load with no additional PV production. The latter corresponds to the minimum daily peak load which cannot be met by the PV output. In (9.9), we considered the effect of short-term battery capacity. For a longer storage, the daily average power of PV in day i is given as:

$$\overline{P}^{sd}(i) = \frac{\sum_{k=i-sd}^{i} \overline{P}^{0}(k)}{sd+1} \tag{9.10}$$

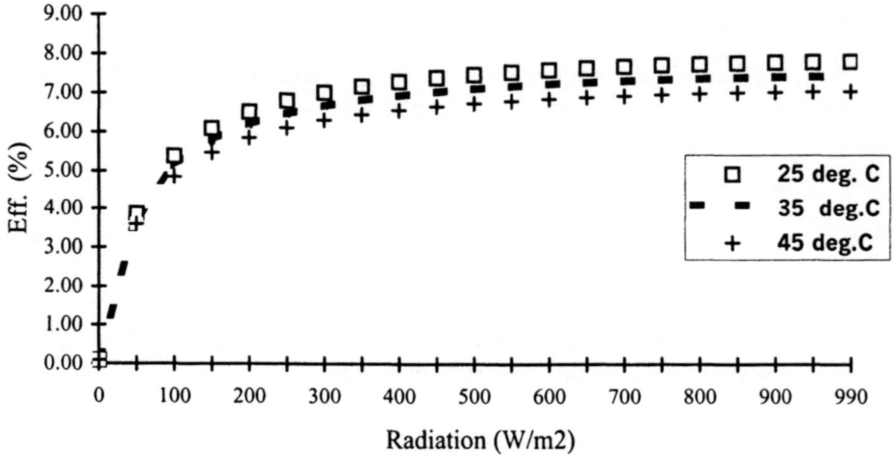

Figure 9.2 PV Efficiency vs. Radiation

9.4 PRODUCTION COST ANALYSIS

The probabilistic modeling in our study simulates the dispatch of generating units and estimates the corresponding production cost. Thermal units are usually represented by a discrete pdf and the PV output is represented by a continuous pdf. In the case of continuous distributions, functions are discretised into a sequence of discrete impulses. If we know the daily average power of PV, then we can prepare the pdf of a PV generator (f_{pv}) within an interval (e.g., 1 year). The f_{pv} as a discrete function with r power impulses of represented by α_m and $\overline{P}^{sd}(m)$ can be described as:

$$f_{pv}(\overline{P}) = \sum_{m=1}^{r} \alpha_m \, \delta(\overline{P} - \overline{P}^{sd}(m)) \tag{9.11}$$

In (9.11), α_m is interpreted as a fraction of an interval in which PV produces the daily average power of $\overline{P}^{sd}(m)$. In general, $\alpha_m=1/Td$ and $r=Td$, where Td is the number of days within the study period. When the number of impulses increase, the sharing process can be utilized [All81]. First we determine the interval:

$$Tp = max\{\overline{P}^{sd}(m)\} - min\{\overline{P}^{sd}(m)\} \tag{9.12}$$

which is divided into a fixed number of points Npv=INT($Tp/\Delta P$). When a power impulse falls between two pre-determined points, it will be shared by both points using an averaging method based on the distance of the impulse from the two fixed points. Using this process, we reduce the number of power impulses to Npv. This process is shown in Figure 9.3.

Now, we treat PV as a multi-state unit with Npv states using the EEF production cost [Wan88]. The operating capacity of each state is $s.\Delta P$ with a probability of $f_{pv}(s.\Delta P)$ where $s=0,1,\ldots,Npv$. Obviously:

$$\sum_{s=0}^{Npv} f_{pv}(s.\Delta P) = 1 \tag{9.13}$$

Let $E(J)$ be the electric energy corresponding to the an interval $[(J-1)\Delta P, J\Delta P]$. From the definition of EEF for a multi-state unit, we find:

$$E^{(i)}(J) = \sum_{s=0}^{Npv} E^{(i-1)}(J-s)\; f_{pv}(\overline{s}.\Delta P) \tag{9.14}$$

The formulation of EEF is outlined in Appendix B. If units are loaded in merit order, then PV will be the first one to be committed, and hence Eq. (9.14) becomes:

$$E^{(1)}(J) = \sum_{s=0}^{Npv} E^{(o)}(J-s)\; f_{pv}(\overline{s}.\Delta P) \tag{9.15}$$

where $E^{(o)}$ is the original EEF of the load and $E^{(1)}$ is the EEF of the load after committing PV. The unserved energy for the load at present is:

$$E_{DI} = \sum_{J=1}^{NE} E^{(o)}(J) - \sum_{s=1}^{Npv} \left[f_{pv}(s.\Delta P) . \sum_{J=1}^{s} E^{(o)}(J) \right] \tag{9.16}$$

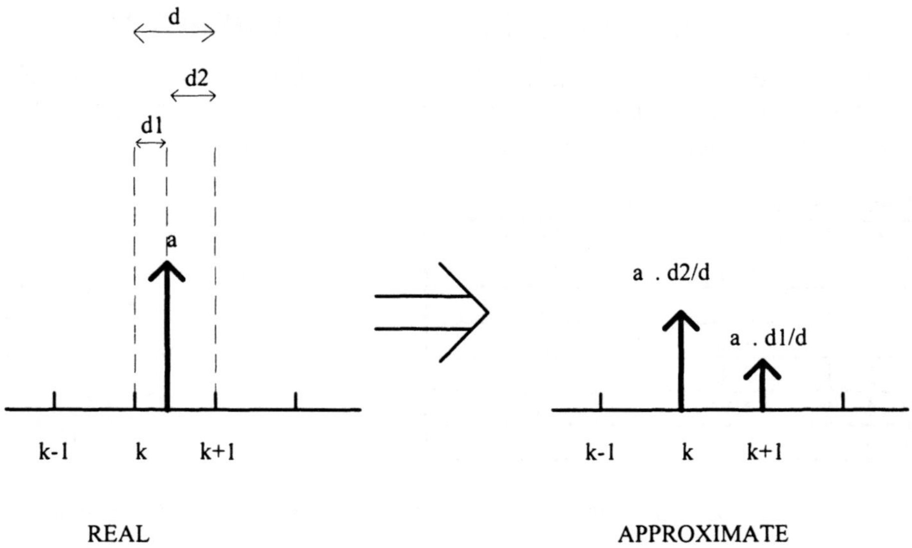

Figure 9.3 Sharing Process of an Impulse

Now, we commit thermal units in merit order. Let q_i be the forced outage rate of thermal unit i with Cp_i capacity. Assume the EEF of the system is $E^{(i-1)}(J)$ after (i-2) thermal units and PV are committed, then the EEF of the load after the (i-1)-th thermal unit is committed is given as:

$$E^{(i)}(J) = p_i E^{(i-1)}(J) + q_i E^{(i-1)}(J - K_i) \tag{9.17}$$

where K_i = INT(Cp_i /ΔP). After the i-th unit is committed, the first i units serve the load from segment J=1 to J=J_i. The unserved energy for a load, once these first i units (PV and i-1 thermal units) are committed, is given in (9.18).

$$E_{Di} = \sum_{J>J_{i-1}} E^{(i-1)}(J) - p_i \sum_{J=J_{i-1}+1}^{J_i} E^{(i-1)}(J) \tag{9.18}$$

The last term in Eq. (9.18) must be the energy generated by unit i (i.e., thermal unit i-1).

$$E_{Gi} = p_i \sum_{J=J_{i-1}+1}^{J_i} E^{(i-1)}(J) \tag{9.19}$$

9.5 NUMERICAL EXAMPLES AND TEST RESULTS

Consider the daily average tilted radiation in a particular site as shown in Table 9.1. Assume that the ambient temperature and wind speed do not vary significantly and there is no storage day (sd=0).

Table 9.1 Radiation and PV Output

day	Radiation W/m^2	PV Output kW
1	216	32
2	135	20
3	270	40
4	107	15
5	68	10

There is a PV plant in the utility system; PV generators have K_c= 0 and η_c= 7.5%. Since K_c=0, the daily average power of PV($\overline{P}^0(i)$) can be easily calculated by multiplying the daily average tilted radiation with η_c and the PV area (A_c). The pdf of PV with 40 kW intervals is discretised into 10 kW increments (ΔP). If an impulse exists between two pre-determined points, e.g. $\alpha(10)$ and $\alpha(20)$, it will be shared by points using the sharing process described earlier. As impulses are spaced equally, the pdf of PV is depicted in Figure 9.4. In this figure,

$\alpha(0) = 0$

$\alpha(10) = 1/5+ (5/10\text{x}1/5) = 0.3$

$\alpha(20) = 1/5+ (5/10\text{x}1/5) = 0.3$

$\alpha(30) = 8/10 \text{ x}1/5 = 0.16$

$\alpha(40) = 1/5+(2/10\text{x}1/5)= 0.24$

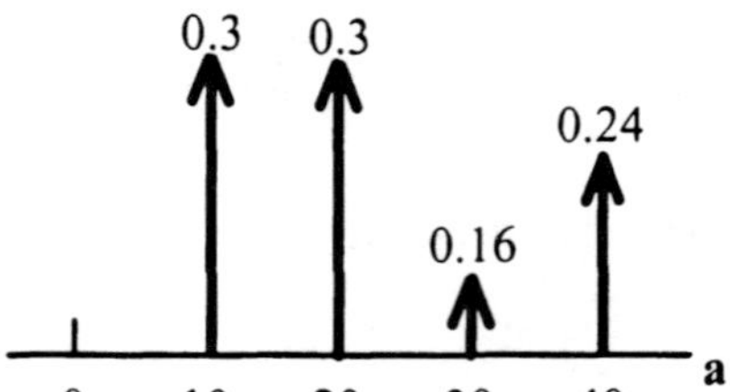

Figure 9.4 Pdf of Photovoltaics

In Figure 9.5, the load curve of the system is given in a chronological form. We calculate EEF for a given load curve, where the total electric energy consumed by the system load is:

1180 +725+600+465+270 =3240 kWh

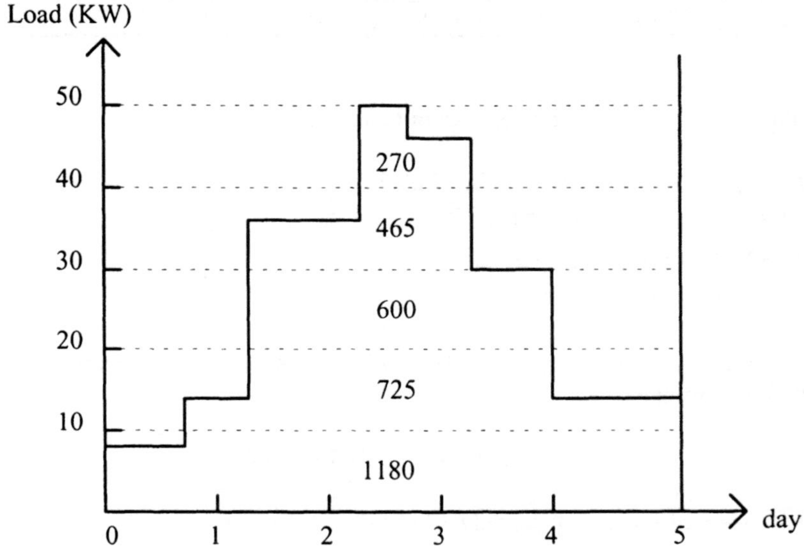

Figure 9.5 Load Curve

The probabilistic modeling for PV generator is given in Table 9.2. The value of EEF after PV is committed is given as:

$E^{(1)}(1)$=725x0.3+600x0.3+465x0.16+270x0.24=536.7 kWh

$E^{(1)}(2)$= 600x0.3+ 465x0.3+270x0.16= 362.7 kWh

$E^{(1)}(3)$= 465x0.3+270x0.3 = 220.5 kWh

$E^{(1)}(4)$= 0.3x270 = 81 kWh

with a total unserved energy of 1200.9 kWh.

Table 9.2 The Probabilistic Modeling for PV

State	*J=1*	*J=2*	*J=3*	*J=4*	*J=5*	*PDF*
$E^{(0)}(J)$	1180	725	600	465	270	0
$E^{(0)}(J\text{-}1)$	725	600	465	270		0.3
$E^{(0)}(J\text{-}2)$	600	465	270			0.3
$E^{(0)}(J\text{-}3)$	465	270				0.16
$E^{(0)}(J\text{-}4)$	270					0.24
$E^{(1)}(J)$	536.7	362.7	220.5	81	0	

The value of EEF after the first thermal unit committed is:

$E^{(2)}(1)$=81x0.9+536.7x0.1= 126.57 kWh
$E^{(2)}(2)$= 362.7x0.1= 36.27 kWh
$E^{(2)}(3)$=220.5x0.1= 22.05 kWh
$E^{(2)}(4)$=81x0.1= 8.1 kWh

with the total unserved energy of 192.99 kWh. The expected generation for the second unit thermal is:

E_{G3}= 0.8x (126.57+36.27+22.05) kWh = 147.912 kWh

The final expected energy not served (EENS) is:

EENS = 192.99-147.912 = 45.078 kWh

In this example, it is apparent that the proposed method can incorporate PV into the EEF simulation, without any modification to the EEF procedure, by treating PV as a multi-state unit. The proposed method is not limited to the EEF and may be incorporated into other probabilistic production cost packages as well [Wan88, All81, Sch84, Rau80, Stre80, Gros88, Boo72].

A program for the proposed method has been employed to model a pilot project for electrification in the Nusa Penida Island (Eastern Islands of Indonesia). The 10 kW PV plant with 200 kWh battery is connected to two 88 kW diesel fuel generators with a 0.1 forced outage rate. The typical climate of the project site is given in Table 9.3, where the PV cell characteristic is the same as that in Figures 9.1 and 9.2. The tilted angle of PV is 5° facing North. The system configuration is given in Figure 9.6, where there are 400 houses (i.e., 2000 population) with a daily consumption of 100 kWh.

Table 9.3 Monthly Climate Data in Nusa Penida Island

Month	Daily horizontal radiation W/m^2	Wind speed m/s	Temp. °C
January	188.33	4.3	26
February	192.50	4.1	28
March	181.66	3.1	27
April	192.92	3.2	27
May	208.75	4.9	31
June	208.75	5.7	30
July	214.16	6.0	32
August	209.58	6.0	29
September	223.33	4.3	25
October	214.16	3.9	29
November	192.50	3.5	27
December	157.92	3.5	26

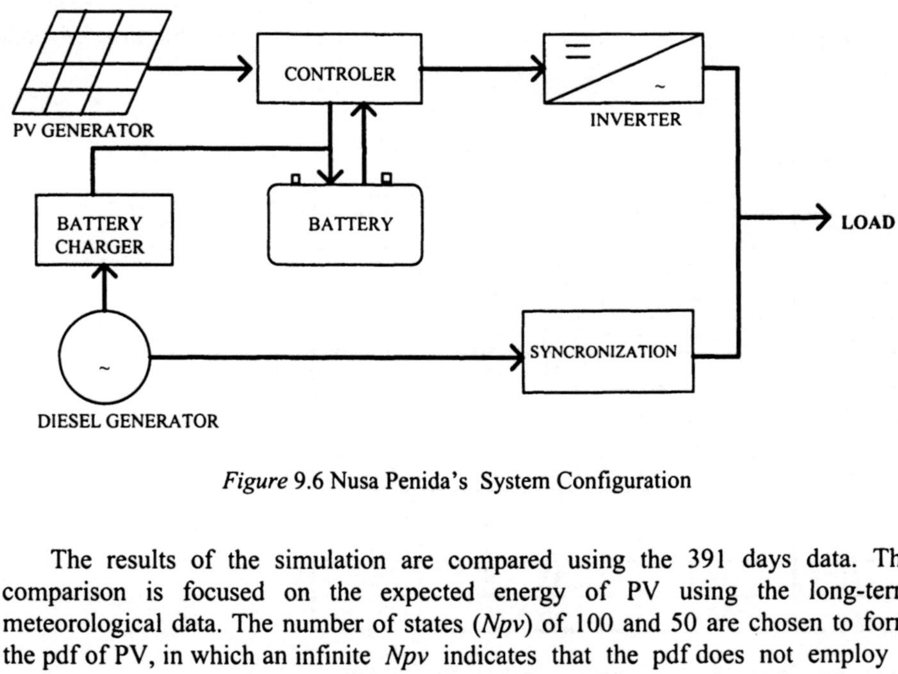

Figure 9.6 Nusa Penida's System Configuration

The results of the simulation are compared using the 391 days data. The comparison is focused on the expected energy of PV using the long-term meteorological data. The number of states (*Npv*) of 100 and 50 are chosen to form the pdf of PV, in which an infinite *Npv* indicates that the pdf does not employ a sharing-process. The results of simulation are summarized in Table 9.4, where the expected energy of PV with different numbers of states is listed. In Table 9.4, it is apparent that the number of states equal to 100 reduces the inaccuracy to 1.76% in the calculation of expected energy; the inaccuracy drops to 3.68% as we reduce the number of states to 50.

Table 9.4 PV Expected Energy Output (MWh)

Month	$Npv = \infty$	Npv=100	Npv=50
1	12.55	12.44	12.18
2	13.25	13.45	13.51
3	11.40	11.26	11.05
4	13.21	13.00	13.62
5	14.28	14.53	14.72
6	14.65	14.42	14.29
7	14.70	14.90	15.17
8	14.40	14.64	14.83
9	15.45	15.65	15.76
10	14.74	14.48	14.45
11	13.03	13.25	13.34
12	10.85	10.64	10.51

The diesel generator's expected energy generation for Npv=100, denoted by E_{G2} and E_{G3}, is summarized in Table 9.5. The output of E_{G2} is larger than that of E_{G3}, because E_{G3} is treated as a back-up unit. E_{G3} is expected to generate about 5.9% of total load, and E_{G2} is expected to generate about 59% of the total load. In practice, we swap the generators every three months to equalize the lifetime of both generators.

Table 9.5 Diesel Generator's Expected Energy Output (MWh)

Month	E_{G2}	E_{G3}	EENS
1	23.99	2.40	0.27
2	23.09	2.31	0.26
3	25.06	2.51	0.28
4	23.49	2.35	0.26
5	22.11	2.21	0.25
6	22.21	2.22	0.25
7	21.78	2.18	0.24
8	22.01	2.20	0.24
9	21.11	2.11	0.23
10	22.16	2.22	0.25
11	23.27	2.33	0.26
12	25.61	2.56	0.28

The effect of PV orientation on the EENS of PV is depicted in Figure 9.7. From March to August, the EENS of 5° tilted angle is less than that of other tilted angles. From September to February 0° tilted angle is more favorable as the site's latitude is 5 degrees on the Southern Hemisphere. If we consider a fixed PV orientation, the minimum EENS of PV can be achieved when PV is tilted 5 degrees facing North.

The effect of the battery size on the EENS of PV is depicted in Figure 9.8. The battery is designed based on a daily output of 40 kWh/day for PV and is set to be used only 60% of its capacity. If we set *sd*=3, the battery capacity is approximated to 200 kWh. For *sd*=5 the battery capacity is approximated to 340 kWh. The effect of *sd* on the PV's EENS is not significant for *sd* >3. If the battery capacity is reduced to 70 kWh (*sd*=1), the PV's EENS will be increased significantly especially during the rainy seasons (October-December, January-March).

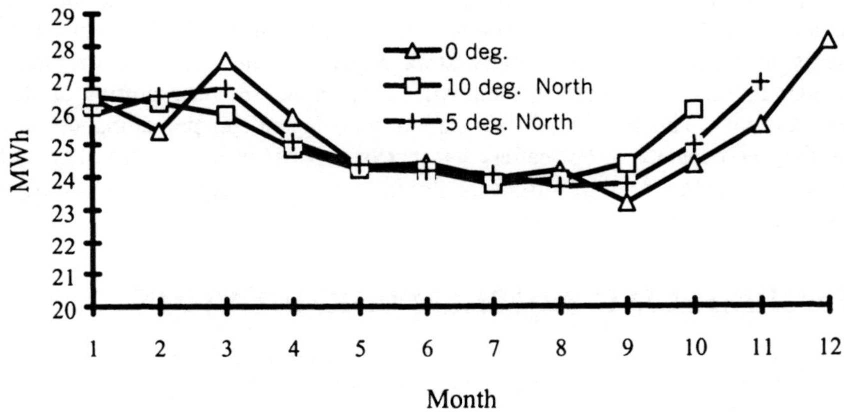

Figure 9.7 Tilted Angle Effect on EENS of PV

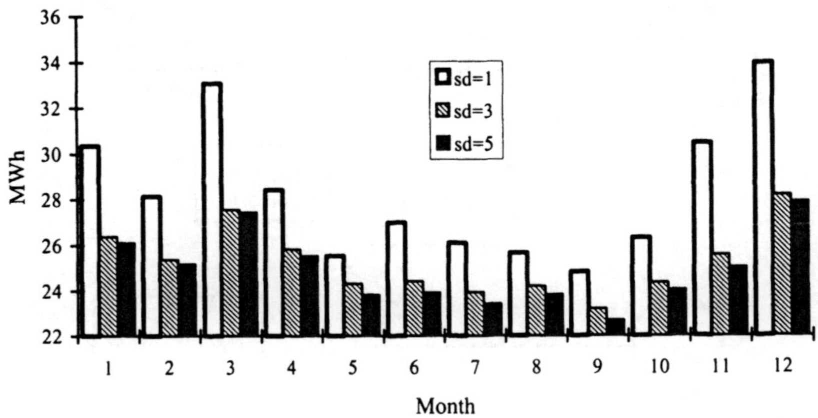

Figure 9.8 Battery Capacity Effect on EENS of PV

9.6 SHORT-TERM GENERATION SCHEDULING IN PHOTOVOLTAIC-UTILITY GRID

The most critical problem in operating a PV plant arises from the fact that renewable sources cannot be dispatched in the same manner as conventional plants. The fluctuations in solar irradiance may occur in a minute-to-minute time frame. Short-term generation changes in an integrated PV-utility, such as those caused by clouds moving over PV systems, the resulting utility load following, and spinning reserve requirements were addressed in [Lee81, Cha85].

It is conceivable that the battery storage is one of the alternatives to the random availability of PV. The results presented in [Cho88, Yam81, Fis79] show that PV systems in conjunction with battery storage play a unique role in demand-side management (peak load shaving), and will likely impact the restructuring of electric power systems. The impact of PV systems will depend on prevailing constraints including the capacity of PV-battery vs. that of thermal utility system. The battery constraints include rate of discharge limit, charge/discharge current limit, and capacity limit at a particular hour.

9.7 SCHEDULING PROBLEM FORMULATION

The PV-utility encompasses the energy generated by PV plants for minimizing the cost of operating thermal units. Figure 9.9 presents an example for a PV-utility system; the PV system may be spread out over a large geographical area with a centralized battery.

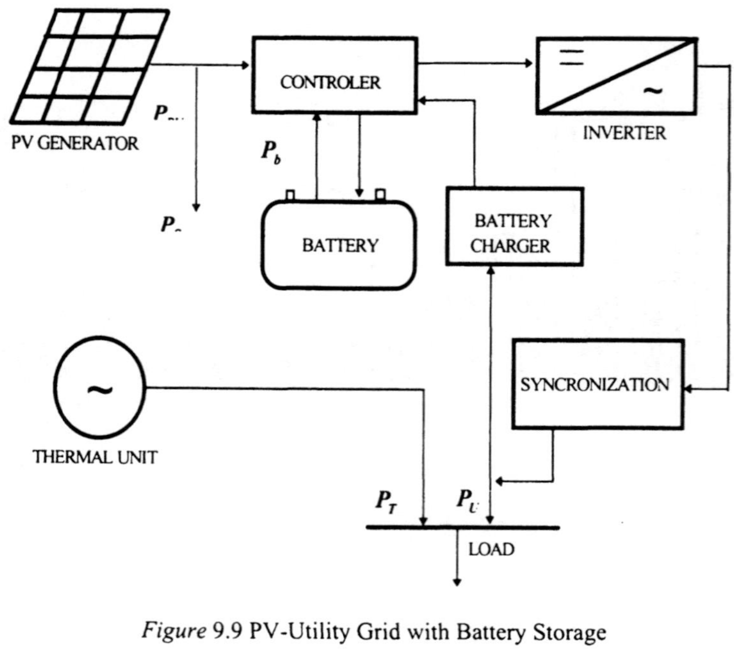

Figure 9.9 PV-Utility Grid with Battery Storage

The mathematical scheduling problem in which the thermal system is represented by an equivalent unit g, with a PV-battery plant P_u is generalized by (9.20). The short-term generation scheduling problem is to determine the hours for which thermal units of an electric power utility should be committed. As discussed in Chapter 4, in order to produce a scheduling procedure that is practical, it is essential that numerous and complex constraints are incorporated into the solution method.

The objective of (9.20) is to minimize the total production over the operational planning period. Constraint (i) points out that the total power generation by thermal units and PV plants should be equal to the load demand. Constraint (ii) represents the charge/discharge equation for the battery. The battery model used in this chapter will be discussed in the next section. Considering the presence of the battery storage, the power from PV cell (P_{pv}) is not the same as that from PV plant (P_u), constraint (iii) represent the equation for PV plant output (P_u).

$$
\begin{aligned}
&\text{Minimize } PC = \sum_{\tau=1}^{N_\tau}\sum_i F(g_{i\tau}) \\
&\text{Subject to} \\
&\sum_i g_{i\tau} + P_u(\tau) = P_D(\tau) && \text{for all } \tau && (i) \\
&C(\tau) = C(\tau-1) + \left[\frac{\Delta\tau\,\eta_b(\tau)}{V_b(\tau)}(P_{pv}(\tau) - P_u(\tau) - P_s(\tau))\right] && \text{for all } \tau && (ii) \\
&P_{pv}(\tau) - P_b(\tau) - P_u(\tau) - P_s(\tau) = 0 && \text{for all } \tau && (iii) \\
&P_{pv}(\tau) = f(G_t(\tau), T_c(\tau)) && \text{for all } \tau && (iv) \\
&P_u(\tau) < \overline{P}_u && \text{penetration limit for all } \tau && (v) \\
&|P_b(\tau)| < \overline{P}_b && \text{charge/discharge limit for all } \tau && (vi) \\
&C(\tau)\Big|_{\tau=0} = C_s && \text{starting state of charge} && (vii) \\
&C(\tau)\Big|_{\tau=N_\tau} = C_f && \text{ending state of charge} && (viii) \\
&\underline{C} < C(\tau) < C_n && \text{state of charge limit for all } t && (ix) \\
&P_b(\tau)\Big|_{t\in tm} = P_f && \text{fixed charge for a particular hour} && (x) \\
&\text{Thermal Plant Constraints such as fuel, generation limits,} && && (xi) \\
&\text{ramp rate, etc.}
\end{aligned}
\qquad (9.20)
$$

The relation among radiation resource, temperature and PV power output is represented by constraint (iv). This relationship is different for each type of PV cells. The relationship among these variables is non-linear, but for practical purposes the linearized relationship can be used. The efficiency of PV for different radiation intensities is depicted in Figure 9.2. At first, a small increase in the radiation produces a significant increase in the PV efficiency. After a certain radiation point, called K_c, further increases in radiation produce relatively small change in the efficiency. Let us denote the corresponding efficiency as η_c, then the instantaneous PV power can be approximated as:

$$P_{pv}(\tau) = \begin{cases} \frac{\eta_c}{K_c}(G_t)^2 & 0 < G_t < K_c \\ \eta_c G_t & G_t > K_c \end{cases} \tag{9.20.a}$$

The intermittent nature of PV adds a significant variance to the thermal generation of a power system. Utilities have traditionally planned for normal load fluctuations and sudden loss of the largest generating unit for secure operation. For the most part, penetration of PV performs well up to about a certain percentage, i.e., 5% of the scheduled load at time [Cha85]. In certain generation dispatches, this penetration can be increased which depends on available units and their capability for performing regulatory duty. This varies based on the season and will be different for each utility. Constraint (v) represents the maximum penetration of PV to the system. Constraints (vi) represent the maximum charge and discharge capacity of the battery. This limit depends on the rating of the battery.

The problem formulation is applicable to other renewable energy sources with intermittent nature. If we consider, for example wind turbine generator instead of PV, $P_{pv}(\tau)$ becomes $P_w(\tau)$ in (ii), (iii), (iv) and $P_w(\tau)$ may be formed as:

$$P_w(\tau) = \begin{cases} 0 & v_w \le v_1 , v_w \ge v_3 \\ \Phi(v_w) & v_1 \le v_w \le v_2 \\ P_n & v_2 \le v_w \le v_3 \end{cases} \tag{9.20.b}$$

where $\Phi(v_w)$ is usually approximated by a linear or quadratic function of or quadratic function of v_w and P_n is the wind generator rated power output, v_1, v_2, v_3 are the wind generator cut in, rated and cut-out wind speeds respectively.

The scheduling problem involves finding the minimum cost trajectory (i.e., the state of charge of the battery at each time period). The charge and discharge of the battery is, of course, fixed by its initial and ending state of charge (SOC). The initial and ending SOC are represented by constraints (vii) and (viii) respectively.

For maintenance reasons, the battery is charged at fixed power for a few hours at particular time to avoid sulfation. This circumstance is represented by constraint (x). Constraints (xi) are thermal plant constraints. These constraints include fuel, generation limit ramp rate limit, environment, minimum up/down time of thermal units, transmission line capacity limits and spinning reserve requirements. The detailed formulation of these constraints is given next.

9.8 PROPOSED METHOD FOR SCHEDULING

Problem (9.20) is a non-liner mixed-integer optimization which consists of integer variables for thermal unit commitment and continuos variables for economic PV-thermal dispatch. We decompose the problem into integer and continuous variable optimizations. We utilize the augmented Lagrangian relaxation to solve integer optimization since it has the capability to overcome the difficulties of complex constraints. To solve the non-linear optimization problem with continuous variables, we utilize successive DP to overcome the difficulties of the long computation time and the large memory storage required by the conventional DP.

To decompose the problem into augmented Lagrangian relaxation and a successive (multi-pass) DP, we employ a three-step formulation. The first step is to set up initial feasible solution for thermal unit commitment. Then we solve the thermal unit commitment problem. The third step is to optimize the PV-thermal dispatch based on thermal unit commitment. The result of this scheduling gives us the loading of each thermal unit. The proposed method is depicted in Figure 9.10.

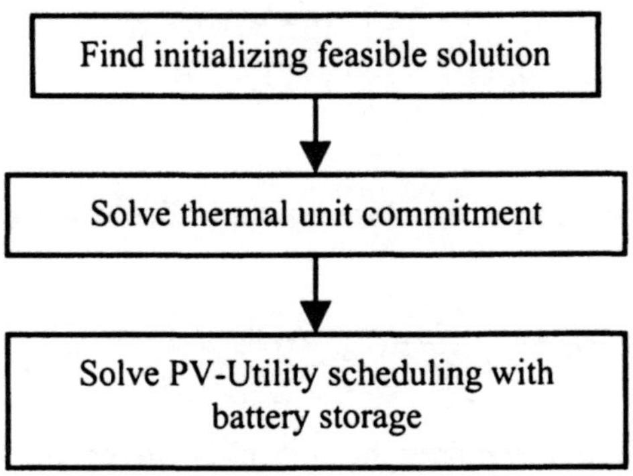

Figure 9.10 Generation Scheduling Flowchart

9.8.1 Initial Feasible Solution. The purpose of this calculation is to produce a possible PV-battery generation in the initial schedule, then the result will be used in the thermal unit commitment calculation. The objective function is to minimize the shortage of energy if there are no thermal units. The battery's SOC levels are used as state variables. Both starting and ending SOC, C_s and C_f, are given along with loading periods. The constraints are available radiation and PV-battery limits. The mathematical model relating radiation and PV power output will be discussed later. The formulation for initial feasible solution is given in (9.21).

$$Minimize\ \ ET = \sum_{\tau=1}^{N_t}(d(\tau)-P_u(\tau))$$

$$\begin{aligned}
&Subject\ to \\
&C(\tau)=C(\tau-1)+\left[\frac{\Delta\tau\,\eta_b(\tau)}{V_b(\tau)}(P_{pv}(\tau)-P_u(\tau)-P_s(\tau))\right] && for\ all\ \tau \\
&P_{pv}(\tau)-P_b(\tau)-P_u(\tau)-P_s(\tau)=0 && for\ all\ \tau \\
&P_{pv}(\tau)=f(G_t(\tau),T_c(\tau)) && for\ all\ \tau \\
&P_u(\tau)<\overline{P}_u && penetration\ limit\ \ for\ all\ \tau \\
&\left|P_b(\tau)\right|<\overline{P}_b && charge/discharge\ limit\ \ for\ all\ \tau \\
&C(\tau)\Big|_{\tau=0}=C_s && starting\ state\ of\ charge \\
&C(\tau)\Big|_{\tau=N_\tau}=C_f && ending\ state\ of\ charge \\
&\underline{C}<C(\tau)<C_n && state\ of\ charge\ limit\ \ for\ all\ \tau \\
&P_b(\tau)\Big|_{t\in tm}=P_f && fixed\ charged\ for\ a\ particular\ hour
\end{aligned}\tag{9.21}$$

If each interval is $\Delta\tau$ hours long, the charging state in battery changes as

$$C(\tau)=C(\tau-1)+\frac{\Delta\tau\eta_b(\tau)}{V_b(\tau)}(P_{pv}(\tau)-P_u(\tau)-P_s(\tau)) \tag{9.22}$$

The second term in (9.22) determines the amount of charge/ discharge of the battery, which is limited to $\overline{P_b}$. If $C(i)$ and $C\ (k)$ denote two different charging states of the battery, the contribution of PV plant to the utility during interval $\Delta\tau$ is:

$$P_u(i,\tau-1:k,\tau)=\frac{(C(i)-C(k))V_b(\tau)}{\eta_b(t)\ \Delta\tau}+P_{pv}(\tau)-P_s(\tau) \tag{9.23}$$

where P_u can be either positive or negative: a positive P_u indicates power flow from the PV plant to the utility.

The initial feasible problem involves finding the minimum shortage of energy trajectory (i.e., charging state at each stage) if there are no thermal units. Numerous feasible trajectories may exist. Let $\{i\}$ and $\{k\}$ be charging states at the beginning and the end of period τ respectively. $ET_k(t)$ is the shortage of energy from the beginning of scheduling period to the end of period t for charging state $C(k)$. The forward DP algorithm is then,

$$ET_k(\tau) = \min_{\{i\}} [ET_i(\tau - 1) + P_D(\tau) - P_u(i, \tau - 1 : k, \tau))] \tag{9.24}$$

The second term in (9.24) is the shortage of power in period t for transition from an initial SOC of $C(i)$ to the end of SOC of $C(k)$. In this case, loads and PV output are given, and the discharge rate of battery is fixed by initial and ending storage levels which in turn establish P_u.

There may well be SOC in the set $C(k)$ that are unreachable by some of initial SOC $C(i)$ because of the constraints in (9.21). A variety of battery constraints and conditions may be incorporated in DP. For example, the battery should be charged at fixed power at a particular hour to avoid sulfation.

We start the search using a coarse grid on both the time interval and the SOC. This permits the refinement of a search for optimal trajectory after a crude search is established in the general neighborhood. Finer grid steps bracketing the range of coarse steps around the initial optimal trajectory could then be used to establish a better path. This method works well for problems with convex objective function. The process is iterative and hence the solution may be local optimum. However, all cases tested in our study converge to reasonable solutions agreeing with engineering intuition.

9.8.2 Thermal Unit Commitment. We apply augmented Lagrangian relaxation [Wan93, Abd96] to solve the thermal unit commitment problem. The production cost $F(g_{i\tau})$ is calculated as the product of the heat rate (MBTU/hr) and the unit's fuel cost ($/MBTU). In some applications, the quadratic cost function is acceptable to describe the input-output characteristics of thermal units, i.e.,

$$F_i(g_{i\tau}) = a_{1i} + a_{2i} g_{i\tau} + a_{3i} g_{i\tau}^2(t) \quad \$/h \tag{9.25}$$

where a_{1i}, a_{2i} and a_{3i} are the coefficients of fuel cost function of fuel thermal i.

Suppose a PV-utility plant with N thermal units. Given $\mathbf{P}_u(\tau)$ by (9.19), the total load of thermal units at each stage is:

$$\sum_i g_{i\tau} = P_D(\tau) - P_u(\tau) \tag{9.26}$$

Then the objective function of thermal unit commitment problem is:

$$PC = Min \sum_i \sum_\tau \left[I_{i\tau} F_i(g_{i\tau}) + S_{i\tau} \right] \tag{9.27}$$

The start up cost of a unit depends on how long the unit was off line and is defined as:

$$S_{i\tau} = I_{i\tau}[1 - I_{i(\tau-1)}]\left[\alpha_i + \beta_i\left(1 - exp\frac{-Z^{off}_{i\tau}}{\Theta_i}\right)\right] \tag{9.28}$$

The constraints of the optimization problem are:

1) System real power balance equation

$$\sum_i I_{i\tau} g_{i\tau} = P_D(\tau) - P_u(\tau) \qquad \tau = 1, \ldots, t \tag{9.29}$$

2) System spinning reserve requirements

$$\sum_i r_{si}(\tau) I_{i\tau} \geq R_s(\tau) \qquad \tau = 1, \ldots, t \tag{9.30}$$

where, $r_{si}(\tau) = min\{10 * MSR_i, \bar{g}_i - g_{i\tau}\}$.

The spinning reserve requirement, $R_s(\tau)$, is typically defined as a base component plus a fraction of the load requirement and a fraction of the high operating limit of the largest on-line unit.

3) System operating reserve requirements

$$\sum_i r_{oi}(\tau) I_{i\tau} \geq R_o(\tau) \qquad \tau = 1, \ldots, t \tag{9.31}$$

where, $r_{oi}(\tau) = \begin{cases} q_i, & if\ unit\ i\ is\ OFF \\ r_{si}(\tau), & if\ unit\ i\ is\ ON \end{cases}$

Interruptible loads are usually added to the operating reserve capacity of units (left hand side of Eq. (9.31)). The operating reserve requirement, $R_o(\tau)$, is commonly defined similar to $R_S(\tau)$ as a function of a base component plus a fraction of the load requirement and a fraction of the high operating limit of the largest on-line unit.

4) Unit generation limits

$$\underline{g}_i \leq g_{i\tau} \leq \bar{g}_i \qquad \tau = 1, \ldots, t \qquad \forall i \tag{9.32}$$

5) Thermal unit minimum starting up/down times

$$(Z^{on}_{i(\tau-1)} - T^{on}_i) * (I_{i(\tau-1)} - I_{i\tau}) \geq 0 \tag{9.33}$$

$$(Z^{off}_{i(\tau-1)} - T^{off}_i)*(I_{i(\tau-1)} - I_{i\tau}) \geq 0 \tag{9.34}$$

6) Ramp rate limits

$$g_{i\tau} - g_{i(\tau-1)} \leq UR_i \quad \textit{as unit i ramps up} \tag{9.35}$$

$$g_{i(\tau-1)} - g_{i\tau} \leq DR_i \quad \textit{as unit i ramps down}$$

7) Fuel constraints

$$FL \leq \sum_{\tau}\sum_{i\in A} f_{ci} * [I_{i\tau}F_i(g_{i\tau}) + S_{i\tau}] \leq FU \tag{9.36}$$

8) Transmission line capacity limits

$$\underline{f}_k \leq f_{k\tau} \leq \overline{f}_k \tag{9.37}$$

$$f_{k\tau} = \sum_{i=1}^{N} A_{mi} g_{i\tau} \boldsymbol{I}_{i\tau} - \sum_{j=1}^{NB} A_{mj}(\boldsymbol{P}_{Dj}(\tau) - \boldsymbol{P}_{uj}(\tau)) \quad m = 1,\ldots,M$$

the sensitivity coefficients A_{mi} and A_{mj} can be determined based on dc or ac power flow corresponding to normal or contingency condition.

9) System emission limit

$$\sum_{\tau}\sum_{i} H_i(g_{i\tau})I_{i\tau} \leq EMS_t \tag{9.38}$$

where several emission types (e.g. SO_2, NO_X) are considered.

10) Area emission limit

$$\sum_{\tau}\sum_{i\in B} H_i(g_{i\tau})I_{i\tau} \leq EMA_t \tag{9.39}$$

where $k \in A$ corresponds to all units in the constrained emission area A; several emission types can be considered.

9.8.3 Dynamic Economic Dispatch. The objective of the dispatch problem is to achieve the minimum production cost by properly scheduling of PV units, thermal units, and battery energy storage units for the given period. The results must meet system constraints. Given the available thermal units in each stage ($I^*_{i\tau}$), we solve the PV-thermal scheduling problem. The formulation is expressed in (9.40).

$$\text{Minimize} \quad PC = \sum_i \sum_\tau I_{i\tau}^* F_i(g_{i\tau})$$

Subject to

$$\sum_i I_{i\tau}^* g_{i\tau} + P_u(\tau) = P_D(\tau) \qquad \text{for all } \tau$$

$$g_{i\tau} < \bar{g}_i \, I_{i\tau}^* \qquad \text{for } i \in N_\tau \text{ and for all } \tau$$

$$C(\tau) = C(\tau-1) + \left[\frac{\Delta\tau \, \eta_b(\tau)}{V_b(\tau)} (P_{pv}(\tau) - P_u(\tau) - P_s(\tau)) \right] \qquad \text{for all } \tau$$

$$P_{pv}(\tau) - P_b(\tau) - P_u(\tau) - P_s(\tau) = 0 \qquad \text{for all } \tau$$

$$P_{pv}(\tau) = f(G_t(\tau), T_c(\tau)) \qquad \text{for all } \tau$$

$$P_u(\tau) < P_u^{max} \qquad \text{penetration limit for all } \tau$$

$$\left| P_b(\tau) \right| < P_b^{max} \qquad \text{charge/discharge limit for all } \tau$$

$$C(\tau)\Big|_{\tau=0} = C_s \qquad \text{starting state of charge}$$

$$C(\tau)\Big|_{\tau=N_\tau} = C_f \qquad \text{ending state of charge}$$

$$\underline{C} < C(\tau) < C_n \qquad \text{state of charge limit for all } \tau$$

$$P_b(\tau)\Big|_{t \in tm} = P_f \qquad \text{fixed charge for a particular hour}$$

(9.40)

Formulation in Eq. (9.40) is the same as that in Eq. (9.21), except thermal units are considered as well in Eq. (9.40). We solve Eq. (9.40) using a successive DP which is the same method used in solving Eq. (9.21). Solution of Eq. (9.40) gives us the updated total load of thermal units:

$$P_T(\tau) = \sum_{i \in N_t} I_{i\tau}^* g_{i\tau} \qquad (9.41)$$

9.9 PV GENERATION PLANT MODEL

For generation scheduling and dispatch, electric utilities are interested in the availability of PV power on an hourly basis. The PV power assessment in electric utilities begins with modeling the solar energy. The hourly meteorological data are needed to simulate the performance of the solar energy. Ideally, many years of data would be necessary for the simulation and the statistical representation of complex meteorological variables.

In practice, long-term data are primarily available in limited sites, and lack of data has prompted the development of a variety of methods to synthesize the existing meteorological data [Liu63, Org14]. Unpredictable surplus or deficit energy from PV can be temporarily controlled by increasing or decreasing PV's spillage power (P_s) or turning on or off the thermal peak-units. In the case that the weather trend is different, a corrective schedule is imposed by re-calculating problem (9.20).

9.9.1 Radiation on Tilted Surface. Long-term meteorological data are usually available in the form of irradiance on horizontal surface. However, in all likelihood, the PV system will employ array tilting or constant tracking of the sun for maximum efficiency. Therefore, irradiance measured or estimated on a horizontal surface must be translated to inclined surfaces using the method in Appendix C.

9.9.2 PV Cell Model. The calculation of the output current of a PV generator is dependent on the PV generator voltage, radiation on the PV generator, wind speed and ambient temperature. The relationship between voltage V_c of a cell of PV and current I_c is represented by a diode model. Details of the model can found in Appendix C.

9.9.3 Battery Model. A battery is treated as constant voltage source. The energy produced by PV is computed first, and load is supplied with the excess energy available to charge the battery. If the system load exceeds the PV output, the difference is drawn from the battery until the battery is fully discharged. This charge accounting scenario may be repeated on an hourly basis. This procedure provides a reasonable approach, however, voltage variations in a battery are required for determining the operating point of the system. Correspondingly, we devise the method in [She65, Bey87] to simulate the battery operation, more accurately.

Needless to say that the actual size of battery will depend on the amount of peak shaving desired. Some utilities may have a load profile, which will not allow peak shaving beyond a certain limit, and constrained by the depth of discharge limitations on the battery itself. A second factor is the fact that the cost of battery is largely dependent on the MWh size of the plant rather than the MW size. Thus utility planners would opt for a low MWh to MW ratio in sizing a battery plant. That means a small period of discharge. Also, the fixation of an optimal amount of peak shaving is the limitation on the total base case capacity available for charging the battery. The best possible use of a PV generation is during the peak shaving period, which decreases the capacity needed from battery discharge during the other hours and is therefore conducive to the battery sizing. Reducing the peak load-shaving amount certainly precludes the PV power from being optimally utilized and therefore works against the economics of the utility.

9.10 CASE STUDIES

A system with 26 thermal units is used to test the proposed method [Wan93]. The hourly radiation data which is used in conjunction with PV is given in Table 9.6. The PV-battery plant penetration is limited to 300 MW. The minimum SOC is limited to 40% and SOC is represented by 20 discrete values. The battery constraints are depicted in Table 9.7. The battery and PV parameters are the same as those in Appendix C. We assume there is no fixed charge required for the battery, and the total PV capacity installed is 260 MWp. There is no spillage power from PV (P_s = 0). Renewable energy components are simulated using the software package INSEL [Bey87].

Table 9.6 Average Hourly Radiation

Hour	G_t (W/m2)	Hour	G_t (W/m2)	Hour	G_t (W/m2)
1	0	9	637.5	17	255.0
2	0	10	680.0	18	212.5
3	0	11	816.0	19	153.0
4	0	12	850.0	20	68.0
5	93.5	13	833.0	21	42.5
6	212.5	14	850.0	22	0
7	255.0	15	680.0	23	0
8	467.5	16	595.0	24	0

Table 9.7 Battery Constraints

MWh min	MWh max	Pbmax MW	SOC initial	SOC final
1400	3500	270	50%	50%

The effect of PV and battery on thermal generation can be seen in Figure 9.11. The presence of PV without battery reduces the thermal generation during hours 5 to 21 which in turn reduces the production cost. The most severe condition created by PV generation without battery is resulted when the radiation decreases but the load increase (hours 17 to 19). This appears to thermal units as a large sudden load change. These large load changes may not be tolerated since thermal unit commitment schedule is limited by ramp rate and minimum up/down time of the thermal units. The battery provides a greater flexibility to schedule thermal units as follows:

1. To avoid the commitment of expensive thermal units during peak load hours, which in turn reduces the fuel cost.
2. To avoid the base generation, such as nuclear plant, to be shut down in low load hours.
3. To avoid the frequent start up and shut down of thermal units, which in turn reduces the start up cost.

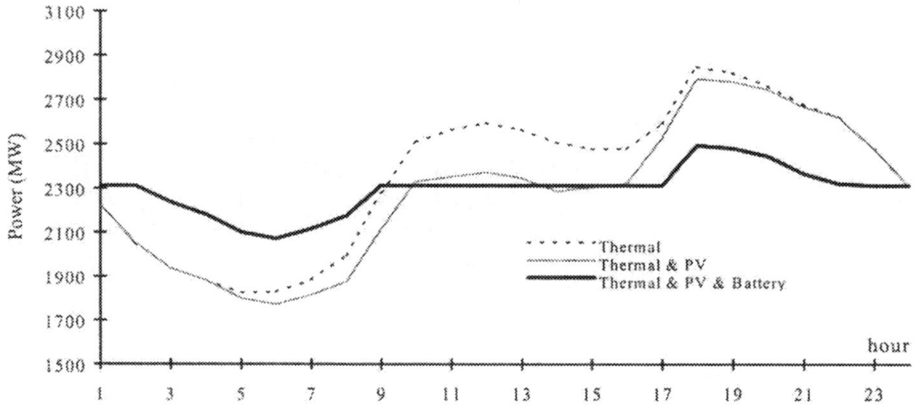

Figure 9.11 Thermal Unit Generation

The production cost saving due to the presence of battery and PV is seen in Table 9.8. We test four different cases. In the first case, the load is supplied by thermal units only. In the second case, we add the battery to the system for the load peak shaving. In the third case, load is supplied by PV and thermal units without battery. In case four, we add the battery to the PV-thermal generation.

From Table 9.8, we see that the PV and battery (case 4) can save a fuel cost of $53,417/day over case 1. We also see that even when there is no PV energy (case 2) the battery provides a saving of $6,610/day over case 1 in the total daily production cost. The battery consumption represents the additional energy needed for charging the battery. For our cases, this consumption is not significant as compared to the benefits of the shaved peak load.

Figure 9.12 shows the penetration of PV plant (P_u). The white block is the PV power flow to utility when we do not use the battery. The dark block is the case as we use PV and battery. At hours 1 to 9, after midnight, the injected power is negative as battery is charged for peak shaving. The battery is designed to save the fuel cost by serving the peak load in the evening (a high fuel-cost load) with stored energy and then charging up at light load periods after midnight (a lower cost load).

Table 9.8 Production Cost

No	Case	Cost ($/day)	Battery Consumption (MWh/day)
1	Thermal only	749,541	-
2	Thermal & Battery	742,931	388
3	Thermal & PV	709,808	-
4	Thermal & PV & Battery	696,124	344

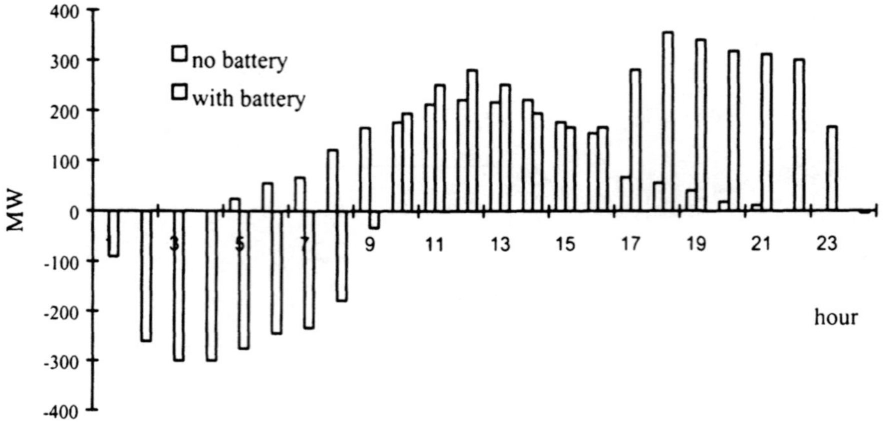

Figure 9.12 Penetration to the Utility from PV Plant

Table 9.9 is the unit commitment without PV penetration; Table 9.10 is the final unit commitment with PV penetration. In neither table, unit commitment considers battery (cases 1 and 3). Comparing Tables 9.9 and 9.10, we learn that there are some differences in scheduling Oil and Coal units (units 12,13,15 and 16) between hours 5 to 18 due to the penetration of PV. During this time, the power of PV is above 50 MW as seen in Figure 9.12 (white blocks). In Table 9.10, units 12,13,15 and 16 can be out for a longer time, but there is no peaking unit advantage. Furthermore, peaking units 1-9 have to be operated more often to compensate for the sudden change in the PV generator output.

If we add the battery (cases 2 and 4), then units which are shown in bold italics will change status, i.e., 0 becomes 1 and 1 becomes 0. The effect of battery can be seen in Table 9.9. Peaking units in case 1 (units 1-9) which have to be started up more than once, operate only at hour 1 in case 2. Unit 10-13, which used to be shutdown at hour 3 and 4, are on at all times in case 2. We learn that the battery reduces the start-up frequency of the thermal unit and reduces the operating time of the peaking unit.

Table 9.9 Thermal Unit Commitment Schedule without PV

Unit	0 1 2 . . . . . hour 24
1	0 1 0 0 ***1*** 0 0 0 0 ***1*** 0 0 0 0 0 0 0 0 ***1*** 0 0 0 0 ***1*** 0
2	0 1 0 0 ***1*** 0 0 0 0 ***1*** 0 0 0 0 0 0 0 0 0 0 0 0 0 0 0
3	0 1 0 0 ***1*** 0 0 0 0 ***1*** 0 0 0 0 0 0 0 0 ***1*** 0 0 0 0 ***1*** 0
4	0 1 0 0 ***1*** 0 0 0 0 ***1*** 0 0 0 0 0 0 0 0 ***1*** 0 0 0 0 ***1*** 0
5	0 1 0 0 ***1*** 0 0 0 0 ***1*** 0 0 0 0 0 0 0 0 0 0 0 0 0 ***1*** 0
6	0 1 0 0 0 0 0 0 0 0 0 0 0 0 0 0 0 0 0 0 0 0 0 0 0
7	0 1 0 0 0 0 0 0 0 0 0 0 0 0 0 0 0 0 0 0 0 0 0 0 0
8	0 1 0 0 0 0 0 0 0 0 0 0 0 0 0 0 0 0 0 0 0 0 0 0 0
9	0 1 0 0 0 0 0 0 0 0 0 0 0 0 0 0 0 0 0 0 0 0 0 0 0
10	1 1 1 ***0*** ***0*** 1 1 1 1 1 1 1 1 1 1 1 1 1 1 1 1 1 1 1 1
11	1 1 1 ***0*** ***0*** 1 1 1 1 1 1 1 1 1 1 1 1 1 1 1 1 1 1 1 1
12	1 1 1 ***0*** ***0*** 1 1 1 1 1 1 1 1 1 1 1 1 1 1 1 1 1 1 1 1
13	1 1 1 ***0*** ***0*** 1 1 1 1 1 1 1 1 1 1 1 1 1 1 1 1 1 1 1 1
14	0 1 1 1 1 0 0 0 1 1 1 1 1 1 1 1 1 1 1 1 1 1 1 1 1
15	0 1 1 1 1 0 0 0 1 1 1 1 1 1 1 1 1 1 1 1 1 1 1 1 1
16	0 1 1 1 1 0 0 0 ***0*** 1 1 1 1 1 1 1 1 1 1 1 1 1 1 1 1
17	1 1 1 1 1 1 1 1 1 1 1 1 1 1 1 1 1 1 1 1 1 1 1 1 1
18	1 1 1 1 1 1 1 1 1 1 1 1 1 1 1 1 1 1 1 1 1 1 1 1 1
19	1 1 1 1 1 1 1 1 1 1 1 1 1 1 1 1 1 1 1 1 1 1 1 1 1
20	1 1 1 1 1 1 1 1 1 1 1 1 1 1 1 1 1 1 1 1 1 1 1 1 1
21	0 0 0 0 0 0 0 0 0 ***1*** ***1*** 1 1 1 1 1 1 1 1 1 1 1 1 1 1
22	***0*** ***0*** ***0*** ***0*** ***0*** 0 0 0 0 0 0 ***1*** ***1*** ***1*** ***1*** ***1*** 1 1 1 1 1 ***1*** ***1*** 0 0
23	***0*** ***0*** ***0*** ***0*** ***0*** ***0*** ***0*** ***0*** ***0*** ***0*** 1 ***1*** ***1*** ***1*** ***1*** ***1*** ***1*** ***1*** ***1*** ***1*** ***1*** ***1*** ***1*** 0 0
24	1 1 1 1 1 1 1 1 1 1 1 1 1 1 1 1 1 1 1 1 1 1 1 1 1
25	1 1 1 1 1 1 1 1 1 1 1 1 1 1 1 1 1 1 1 1 1 1 1 1 1
26	1 1 1 1 1 1 1 1 1 1 1 1 1 1 1 1 1 1 1 1 1 1 1 1 1

Table 9.10 Thermal Unit Commitment Schedule with PV

Unit	0 1 2 . . . . . hour 24
1	0 1 0 0 ***1 1*** 0 0 0 0 0 0 0 0 ***1 1 1*** 0 0 0 0 0 0 ***1*** 0
2	0 1 0 0 ***1 1*** 0 0 0 0 0 0 0 0 ***1 1*** 0 0 0 0 0 0 0 0 0
3	0 1 0 0 ***1 1*** 0 0 0 0 0 0 0 0 ***1 1 1*** 0 0 0 0 0 0 ***1*** 0
4	0 1 0 0 ***1 1*** 0 0 0 0 0 0 0 0 ***1 1 1*** 0 0 0 0 0 0 ***1*** 0
5	0 1 0 0 ***1 1*** 0 0 0 0 0 0 0 0 ***1 1 1*** 0 0 0 0 0 0 ***1*** 0
6	0 1 0 0 0 0 0 0 0 0 0 0 0 0 0 0 0 0 0 0 0 0 0 0 0
7	0 1 0 0 0 0 0 0 0 0 0 0 0 0 0 0 0 0 0 0 0 0 0 0 0
8	0 1 0 0 0 0 0 0 0 0 0 0 0 0 0 0 0 0 0 0 0 0 0 0 0
9	0 0 0 0 0 0 0 0 0 0 0 0 0 0 0 0 0 0 0 0 0 0 0 0 0
10	1 1 1 ***0 0*** 1 1 1 1 1 1 1 1 1 1 1 1 1 1 1 1 1 1 1 1
11	1 1 1 ***0 0*** 1 1 1 1 1 1 1 1 1 1 1 1 1 1 1 1 1 1 1 1
12	1 1 1 ***0 0 0 0 0*** 1 1 1 1 1 1 1 1 1 1 1 1 1 1 1 1 1
13	1 1 1 ***0 0 0 0 0*** 1 1 1 1 1 1 1 1 1 1 1 1 1 1 1 1 1
14	0 1 1 1 1 1 0 0 0 1 1 1 1 1 0 0 1 1 1 1 1 1 1 1 1
15	0 1 1 1 1 ***1*** 0 0 0 ***0 0 0 0 0 0*** 0 0 1 1 1 1 1 1 1 1
16	0 1 1 1 1 0 0 0 ***0 0 0 0 0 0 0*** 0 0 1 1 1 1 1 1 1 1
17	1 1 1 1 1 1 1 1 1 1 1 1 1 1 1 1 1 1 1 1 1 1 1 1 1
18	1 1 1 1 1 1 1 1 1 1 1 1 1 1 1 1 1 1 1 1 1 1 1 1 1
19	1 1 1 1 1 1 1 1 1 1 1 1 1 1 1 1 1 1 1 1 1 1 1 1 1
20	1 1 1 1 1 1 1 1 1 1 1 1 1 1 1 1 1 1 1 1 1 1 1 1 1
21	0 ***0 0 0 0 0*** 0 0 0 ***1*** 1 1 1 1 1 1 1 1 1 1 1 1 1 1 1
22	0 0 0 0 0 0 0 0 0 ***1 1 1 1 1 1 1 1 1 1 1 1 1 1*** 0 0
23	0 0 0 0 0 0 0 0 0 0 0 ***1 1 1 1 1 1 1 1 1 1 1 1*** 0 0
24	1 1 1 1 1 1 1 1 1 1 1 1 1 1 1 1 1 1 1 1 1 1 1 1 1
25	1 1 1 1 1 1 1 1 1 1 1 1 1 1 1 1 1 1 1 1 1 1 1 1 1
26	1 1 1 1 1 1 1 1 1 1 1 1 1 1 1 1 1 1 1 1 1 1 1 1 1

APPENDIX A

IEEE-RTS TEST SYSTEM DATA

A.1 Load Model

The annual peak load for the test system is 2850 MW. Table A.1 gives data on weekly peak loads in percentage of annual peak load. The annual peak load occurs in week 51. The data in Table 1 shows a typical pattern with two seasonal peaks. The second peak is in week 23 (90%) with valleys at about 70% in between each peak.

Table A.1 Weekly Peak Load in Percent of Annual Peak

Week	Peak Load	Week	Peak Load
1	86.2	27	75.5
2	90.0	28	81.6
3	87.8	29	80.1
4	83.4	30	88.0
5	88.0	31	72.2
6	84.1	32	77.6
7	83.2	33	80.0
8	80.6	34	72.9
9	74.0	35	72.6
10	73.7	36	70.5
11	71.5	37	78.0
12	72.7	38	69.5
13	70.4	39	72.4
14	75.0	40	72.4
15	72.1	41	74.3
16	80.0	42	74.4
17	75.4	43	80.0
18	83.7	44	88.1
19	87.0	45	88.5
20	88.0	46	90.9
21	85.6	47	94.0
22	81.1	48	89.0
23	90.0	49	94.2
24	88.7	50	97.0
25	89.6	51	100.0
26	86.1	52	95.5

Table A.2 Daily Peak Load in Percent of Weekly Peak

Day	Peak Load
Monday	93
Tuesday	100
Wednesday	98
Thursday	96
Friday	94
Saturday	77
Sunday	75

Table A.3 Hourly Peak Load in Percent of Daily Peak

Hour	Winter Weeks 1-8 &44-52 Wkdy	Winter Weeks 1-8 &44-52 Wknd	Summer Weeks 18-30 Wkdy	Summer Weeks 18-30 Wknd	Spring / Fall Weeks 9-17 &31-43 Wkdy	Spring / Fall Weeks 9-17 &31-43 Wknd
12-1am	67	78	64	74	63	75
1-2	63	72	60	70	62	73
2-3	60	68	58	66	60	69
3-4	59	66	56	65	58	66
4-5	59	64	56	64	59	65
5-6	60	65	58	62	65	65
6-7	74	66	64	62	72	68
7-8	86	70	76	66	85	74
8-9	95	80	87	81	95	83
9-10	96	88	95	86	99	89
10-11	96	90	99	91	100	92
11-Noon	95	91	100	93	99	94
Noon-1pm	95	90	99	93	93	91
1-2	95	88	100	92	92	90
2-3	93	87	100	91	90	90
3-4	94	87	97	91	88	86
4-5	99	91	96	92	90	85
5-6	100	100	96	94	92	88
6-7	100	99	93	95	96	92
7-8	96	97	92	95	98	100
8-9	91	94	92	100	96	97
9-10	83	92	93	93	90	95
10-11	73	87	87	88	80	90
11-12	63	81	72	80	70	85

A.2 Generating System

Table A.4 gives a list of generating unit ratings and reliability data. Table A.5 gives operating and maintenance cost of generating units. For power production, data is given in term of heat rate. Generating unit costs are listed in Table A.6. The locations of generating units are shown in Table A.7. Table A.8 gives data on generating unit MVAr capability for load flow calculations.

Table A.4 Generating Unit Reliability Data

Unit Size MW	Number of Units	Forced Outage Rate	Scheduled Maintenance Wks/year
12	5	0.02	2
20	4	0.10	2
76	4	0.02	3
100	6	0.04	3
155	4	0.04	4
197	6	0.05	4
350	1	0.08	5
400	2	0.12	6

The following fuel costs are used:

#6 oil	$2.30/Mbtu
# 2 oil	$3.00/Mbtu
coal	$1.20/Mbtu
nuclear	$0.60/Mbtu

Table A.5 Generating Unit Operating Cost Data

Size MW	Fuel	Heat Rate Btu/kWh	O&M Costs		Maintenance	
			Fixed $/kW/YR	Variable $/MWh	Window (week)	Duration (week)
12	Oil #6	12000	10.0	5.0	N/A	N/A
20	Oil #2	14500	0.3	5.0	18-29	2
76	Coal	12000	10.0	0.9	18-29	2
100	Oil #6	10000	8.5	0.8	18-29	2
155	Coal	9700	7.0	0.8	N/A	N/A
197	Oil #6	9600	5.0	0.7	N/A	N/A
350	Coal	9500	4.5	0.7	N/A	N/A
400	Nuclear	10000	5.0	0.3	N/A	N/A

Table A.6 Generating Unit Data

No	P^{max}	P^{min}	a_i	b_i	c_i	$1/f_c$	$S_i(t)$
1	12.	2.40	0.02533	25.5472	24.3891	1.0	0.
2	12.	2.40	0.02649	25.6753	24.4110	1.0	0.
3	12.	2.40	0.02801	25.8027	24.6382	1.0	0.
4	12.	2.40	0.02842	25.9318	24.7605	1.0	0.
5	12.	2.40	0.02855	26.0611	24.8882	1.0	0.
6	20.	4.00	0.01561	37.9637	118.9083	1.0	30.
7	20.	4.00	0.01359	37.7770	118.4576	1.0	30.
8	20.	4.00	0.01161	37.9637	118.9083	1.0	30.
9	20.	4.00	0.01059	38.7770	119.4576	1.0	30.
10	76.	15.20	0.00962	13.5073	81.8259	1.0	80.
11	76.	15.20	0.00876	13.3272	81.1364	1.0	80.
12	76.	15.20	0.00895	13.3538	81.2980	1.0	80.
13	76.	15.20	0.00932	13.4073	81.6259	1.0	80.
14	100.	25.00	0.00623	18.0000	217.8952	1.0	100.
15	100.	25.00	0.00599	18.6000	219.7752	1.0	100.
16	100.	25.00	0.00612	18.1000	218.3350	1.0	100.
17	100.	25.00	0.00588	18.2800	216.7752	1.0	100.
18	100.	25.00	0.00598	18.2000	218.7752	1.0	100.
19	100.	25.00	0.00578	17.2800	216.7752	1.0	100.
20	155.	54.25	0.00481	10.7367	142.7348	1.0	200.
21	155.	54.25	0.00473	10.7154	143.0288	1.0	200.
22	155.	54.25	0.00481	10.7367	143.3179	1.0	200.
23	155.	54.25	0.00487	10.7583	143.5972	1.0	200.
24	197.	68.95	0.00259	23.0000	259.1310	1.0	300.
25	197.	68.95	0.00260	23.1000	259.6490	1.0	300.
26	197.	68.95	0.00263	23.2000	260.1760	1.0	300.
27	197.	68.95	0.00264	23.4000	260.5760	1.0	300.
28	197.	68.95	0.00267	23.5000	261.1760	1.0	300.
29	197.	68.95	0.00261	23.0400	260.0760	1.0	300.
30	350.	140.00	0.00150	10.8416	176.0575	1.0	500.
31	400.	100.00	0.00194	7.4921	310.0021	1.0	800.
32	400.	100.00	0.00195	7.5031	311.9102	1.0	800.

Table A.7 Generating Unit Operating Data

No	*Up*	*Dn*	*Ini*	*UR*	*DR*	Bus
1	1	-1	-1	12.0	12.0	15
2	1	-1	-1	12.0	12.0	15
3	1	-1	-1	12.0	12.0	15
4	1	-1	-1	12.0	12.0	15
5	1	-1	-1	12.0	12.0	15
6	1	-1	-1	20.0	20.0	1
7	1	-1	-1	20.0	20.0	1
8	1	-1	-1	20.0	20.0	2
9	1	-1	-1	20.0	20.0	2
10	3	-2	3	38.0	38.0	1
11	3	-2	3	38.0	38.0	1
12	3	-2	3	38.0	38.0	2
13	3	-2	3	38.0	38.0	2
14	4	-2	5	50.0	50.0	7
15	4	-2	5	50.0	50.0	7
16	4	-2	5	50.0	50.0	7
17	4	-2	-3	50.0	50.0	7
18	4	-2	-3	50.0	50.0	7
19	4	-2	-3	50.0	50.0	7
20	4	-2	-3	50.0	50.0	7
21	5	-3	5	77.5	77.5	15
22	5	-3	5	77.5	77.5	16
23	5	-3	5	77.5	77.5	23
24	5	-4	-4	98.5	98.5	22
25	5	-4	-4	98.5	98.5	22
26	5	-4	-4	98.5	98.5	22
27	5	-4	-4	98.5	98.5	22
28	5	-4	-4	98.5	98.5	22
29	5	-4	-4	98.5	98.5	22
30	8	-5	10	175.0	175.0	23
31	8	-5	10	200.0	200.0	21
32	8	-5	10	200.0	200.0	18

Table A.8 Generating Unit MVAr Capability

Size MW	MV Minimum	VAr Maximum
12	0	6
20	0	10
50	-10	16
76	-25	30
100	0	60
155	-50	80
197	0	80
350	-25	150
400	-50	200

A.3 Transmission System

The transmission network consists of 24 buses connected by 38 lines. Bus load data at the time of system peak is shown in Table A.9. Bus loads are assumed to have the same proportional relation to system load as that of peak load conditions. The per unit bus loads are given in the last column of Table A.9. For MVAr requirements, a 98% power factor is assumed. This corresponds to MVAr requirements of approximately 20% of the MW load at each bus. The 98% power factor is assumed to apply at all load levels. Impedance and rating data for lines is given in Table A.10. Transmission line length and forced outage data are given in Table A.11.

Table A.9 Bus Load Data

Bus	Load		Bus Load (% of System Load)
	MW	MVAr	
1	108	22	3.8
2	97	20	3.4
3	180	37	6.3
4	74	15	2.6
5	71	14	2.5
6	136	28	4.8
7	125	25	4.4
8	171	35	6.0
9	175	36	6.1
10	195	40	6.8
13	265	54	9.3
14	194	39	6.8
15	317	64	11.1
16	100	20	3.5
18	333	68	11.7
19	181	37	6.4
20	128	26	4.5

Table A.10 Line Impedance and Rating Data

From Bus	To Bus	p.u. Impedance /100 MVA		Base B	No. of Lines	Rating (MVA)
		R	X			
1	2	.0026	.0139	.4611	1	193
1	3	.0546	.2112	.0572	1	208
1	5	.0218	.0845	.0229	1	208
2	4	.0328	.1267	.0343	1	208
2	6	.0497	.1920	.0520	1	208
3	9	.0308	.1190	.0322	1	208
3	24	.0023	.0839		1	510
4	9	.0268	.1037	.0281	1	208
5	10	.0228	.0883	.0239	1	208
6	10	.0139	.0605	2.459	1	193
7	8	.0159	.0614	.0166	1	208
8	9	.0427	.1651	.0447	1	208
8	10	.0427	.1651	.0447	1	208
9	11	.0023	.0839		1	510
9	12	.0023	.0839		1	510
10	11	.0023	.0839		1	510
10	12	.0023	.0839		1	510
11	13	.0061	.0476	.0999	1	600
11	14	.0054	.0418	.0879	1	600
12	13	.0061	.0476	.0999	1	600
12	23	.0124	.0966	.2030	1	600
13	23	.0111	.0865	.1818	1	600
14	16	.0050	.0389	.0818	1	600
15	16	.0022	.0173	.0364	1	600
15	21	.0063	.0490	.1030	2	600
15	24	.0067	.0519	.1091	1	600
16	17	.0033	.0259	.0545	1	600
16	19	.0030	.0231	.0485	1	600
17	22	.0135	.1053	.2212	1	600
18	21	.0033	.0259	.0545	2	600
19	20	.0051	.0396	.0833	2	600
20	23	.0028	.0216	.0455	2	600
21	22	.0087	.0678	.1424	1	600

Table A.11 Transmission Line Length and Forced Outage

From Bus	To Bus	Length miles	Permanent* Outage Rate	
			Outage Rate	Outage Duration
			1/year	hours
1	2	3	.24	16
1	3	55	.51	10
1	5	22	.33	10
2	4	33	.39	10
2	6	50	.48	10
3	9	31	.38	10
3	24	0	.02	768
4	9	27	.36	10
5	10	23	.34	10
6	10	16	.33	10
7	8	16	.30	768
8	9	43	.44	768
8	10	43	.44	768
9	11	0	.02	11
9	12	0	.02	11
10	11	0	.02	11
10	12	0	.02	11
11	13	33	.40	11
11	14	29	.39	11
12	13	33	.40	11
12	23	67	.52	11
13	23	60	.49	11
14	16	27	.38	11
15	16	12	.33	11
15	21	34	.41	11
15	24	36	.41	11
16	17	18	.35	11
16	19	16	.34	11
17	22	73	.54	11
18	21	18	.35	11
19	20	27.5	.38	11
20	23	15	.34	11
21	22	47	.45	11

*Forced outage rates are calculated as: $FOR = \dfrac{\text{Outage Duration}}{1 / \text{Outage Rate} + \text{Outage Duration}}$

APPENDIX B

PRODUCTION COST MODEL

B.1 Equivalent Energy Function

We present a brief description of equivalent energy function (EEF) in probabilistic modeling of power systems. A detailed treatment is found in [Wan88]. EEF is the energy function that considers failure effects of committed units. Suppose the original load duration curve is described by $f^{(o)}(x)$, the corresponding energy function is $E^{(o)}(J)$

$$E^{(o)}(J) = T\int_{x}^{x+\Delta x} f^{(o)}(x)\,dx \tag{B.1}$$

where J= INT($X/\Delta x$)+1 is an integer. $E(J)$ is the electric energy which corresponds to the area of interval (x, $x+\Delta x$) under the load duration curve.

After the (i-1)th unit is committed, the equivalent load duration curve (ELDC) becomes $f^{(i-1)}(x)$ and EEF becomes $E^{(i-1)}(J)$. If we commit the i-th unit whose capacity is c_i and its FOR is q_i, according to the convolution formula we have

$$f^{(i)}(x) = p_i\, f^{(i-1)}(x) + q_i\, f^{(i-1)}(x-c_i) \tag{B.2}$$

where $p_i = 1 - q_i$. From Eq. (B.1), we find

$$\begin{aligned} E^{(i)}(J) &= T\int_{x}^{x+\Delta x} f^{(i)}(x)\,dx \\ &= T\int_{x}^{x+\Delta x} [\,p_i\, f^{(i-1)}(x) + q_i\, f^{(i-1)}(x-c_i)]\,dx \end{aligned} \tag{B.3}$$

$$E^{(i)}(J) = p_i\, E^{(i-1)}(J) + q_i\, E^{(i-1)}(J-K_i)$$

where K_i= INT ($c_i/\Delta x$).

After the ith unit is committed, the first i units serve the load from segment J=1 to J-1. The unserved energy for the system at present is:

$$E_{Di} = \sum_{J>J_i} E^{(i)}(J) \tag{B.4}$$

Substituting .(B.3) into (B.4)

$$
\begin{aligned}
E_{Di} &= \sum_{J>J_i} [\, p_i \boldsymbol{E}^{(i-1)}(J) + q_i \boldsymbol{E}^{(i-1)}(J-K_i)] \\
&= p_i \sum_{J>J_i} \boldsymbol{E}^{(i-1)}(J) + q_i \sum_{J>J_i} \boldsymbol{E}^{(i-1)}(J-K_i) \\
&= p_i \sum_{J>J_{i-1}} \boldsymbol{E}^{(i-1)}(J) - p_i \sum_{J=J_{i-1}+1}^{J_i} \boldsymbol{E}^{(i-1)}(J) + q_i \sum_{J>J_{i-1}} \boldsymbol{E}^{(i-1)}(J-K_i) \\
&= \sum_{J>J_{i-1}} \boldsymbol{E}^{(i-1)}(J) - p_i \sum_{J=J_{i-1}+1}^{J_i} \boldsymbol{E}^{(i-1)}(J)
\end{aligned} \tag{B.5}
$$

From (B.4) we have

$$
E_{D\,i\text{-}1} = \sum_{J>J_{i-1}} \boldsymbol{E}^{(i\text{-}1)}(J)
$$

Hence, the last term in (B.5) must be the energy generated by unit i,

$$
E_{Gi} = p_i \sum_{J=J_{i-1}+1}^{J_i} \boldsymbol{E}^{(i-1)}(J) \tag{B.6}
$$

and we can write (B.5) as:

$$
E_{D\,i} = E_{D\,i\text{-}1} - E_{G\,i} \tag{B.7}
$$

B.2 Cumulant Method

Details of the cumulant method are given as follows. Let

LK_k	be the k-th cumulant of the hourly load distribution (which can also describe the load duration curve)
$GK_k(r)$	be the k-th cumulant of the generation outage distribution after the r-th unit has been convolved
g_{ki}	the k-th cumulant of the i-th generation unit
p_{ij}	be the probability that the i-th generation unit is in derating state j
O_{ij}	be the outage level of the i-th generation unit in derating state j
L_t	be the hourly load at time t

LM_k be the k-th moment of the load distribution
$ELDC$ Equivalent Load Duration Curve
$ELK_k(r)$ the k-th cumulant of the ELDC after r units have been convolved
M_{ki} be the k-th moment of the outage distribution of the i-th unit in which

$$M_{ki} = \sum_{j=1}^{m} p_{ij} O_{ij}^{k}$$

i (1, . . . ,n), *number of units or capacity blocks*
j (1, . . . , m), *number of outage states*
k (1, 2, 3, 4), *number of cumulants*
t (1, . . . ,T), *number of hours in period*

The cumulants for each unit which determine the generation outage distribution for the system are:

$$\begin{aligned} g_{1i} &= M_{1i} \\ g_{2i} &= M_{2i} - M_{1i}^2 \\ g_{3i} &= M_{3i} - 3M_{2i}M_{1i} + 2M_{1i}^3 \\ g_{4i} &= M_{4i} + 6M_{2i}M_{1i}^2 - 4M_{3i}M_{1i} - 3M_{1i}^4 - 3g_{2i}^2 \end{aligned} \qquad \text{(B.8-B.11)}$$

Similarly, the cumulants of the hourly load distribution are:

$$LM_j = \frac{1}{T}\sum_{t=1}^{T} L_t^j \qquad \textit{for } j = 1, 2, 3, 4 \qquad \text{(B.12)}$$

$$\begin{aligned} LK_1 &= LM_1 \\ LK_2 &= LM_2 - LM_1^2 \\ LK_3 &= LM_3 - 3LM_2LM_1 + 2LM_1^3 \\ LK_4 &= LM_4 + 6LM_2LM_1^2 - 4LM_3LM_1 - 3LM_1^4 - 3LK_2^2 \end{aligned} \qquad \text{(B.13-B.16)}$$

The above calculations are an overhead investment since they will not be revised unless the basic outage data or loads are changed.

In the following, we present the necessary calculations to perform the Gram-Charlier approximation of the equivalent load curve. The cumulants for the equivalent load curve after the first r units have been loaded on the system are:

$$ELK_k(r) = LK_k + GK_k(r) = LK_k + \sum_{i=1}^{r} g_{ki} \qquad \textit{for } k = 1, 2, 3, 4 \qquad \text{(B.17)}$$

and the coefficient of the Gram-Charlier expansion (which is similar to a Taylor series expansion of the equivalent load distribution) are:

$$\mu = ELK_1(r) \qquad \text{(mean)} \tag{B.18}$$

$$\sigma^2 = ELK_2(r) \qquad \text{(variance)} \tag{B.19}$$

$$G_1 = \frac{1}{\sigma^3} ELK_3(r) \qquad \text{(standardized skewness)} \tag{B.20}$$

$$G_2 = \frac{1}{\sigma^4} ELK_4(r) \qquad \text{(standardized kurtosis)} \tag{B.21}$$

Any point along the equivalent load curve facing the r+1 unit can be determined by the following procedure:

Let:

$$Z_1 = \frac{x - \mu}{\sigma} \tag{B.22}$$

$$N(Z) = \frac{1}{\sqrt{2\pi}} exp(\frac{-Z^2}{2}) \tag{B.23}$$

where:

x is the equivalent load level for which we need to know the value of the equivalent load curve i.e., the capacity of the first r units

Z_1 is a standardized variable

$N(Z)$ is the normal distribution evaluated at Z

The Gram-Charlier representation of the Equivalent Load Curve is:

$$ELDC_r(Z_1) = 1 - \int_{\infty}^{Z_1} N(z)dz + \frac{G_1}{3!} N^{(2)}(Z_1) - \frac{G_2}{4!} N^{(3)}(Z_1) - \frac{10}{6!} G_1^2 N^{(5)}(Z_l) \tag{B.24}$$

where, the derivatives of the Normal are:

$$N^{(2)}(Z_1) = (Z_1^2 - 1) N(Z_1) \tag{B.25}$$

$$N^{(3)}(Z_1) = (-Z_1^3 + 3Z_1) N(Z_1) \tag{B.26}$$

$$N^{(5)}(Z_1) = (-Z_1^5 + 10Z_1^3 - 15Z_1) N(Z_1) \tag{B.27}$$

This procedure allows the determination of *ELDC* at a position in the loading order, without having to numerically convolve the previous units.

Example B.1

The load cumulants for the first week's hourly loads have been determined from (B.12) through (B.16) to be:

LK_1= 6000 MW (mean)
LK_2= 4.0x10^6 (variance)
LK_3=-2.4x10^9 (skewness)
LK_4= 9.6x10^{11} (kurtosis)
Peak Load = 4000 MW

The first three generating units in the loading order are 1000 MW each with availabilities of 0.9, 0.8 and 0.7. What is the dual multiplier π_{4t} of the fourth unit, which is 800 MW with availability of 0.9?

Table B.1 Outage Cumulants for the Four Units

	Unit 1	Unit 2	Unit 3	Unit 4
Capacity	1000	1000	1000	800
g_{1i} (mean)	100	200	300	80
g_{2i} (variance)	9.0×10^4	1.6×10^5	2.1×10^5	5.8×10^4
g_{3i} (skewness)	7.2×10^7	9.6×10^7	8.4×10^7	3.7×10^7
g_{4i} (kurtosis)	4.1×10^{10}	6.4×10^9	-5.5×10^{10}	1.7×10^{10}

The Gram-Charlier parameters for the *ELDC* facing the fourth plant in the loading order are calculated from (B.17) to be:

$ELK_1(3)$=6000+100+200+300=6600
$ELK_2(3)=4.5 \times 10^6$
$ELK_3(3)=-2.1 \times 10^9$
$ELK_4(3))=9.5 \times 10^{11}$

And from (B.18) through (B.21), we have:

μ=6600
σ^2=4.5x106
G_1=-0.22
G_2=0.047

Now that the coefficients of approximation to the equivalent load curve have been determined, we evaluate the approximation loading point (L) at L_3=3000 and L_4=3800.

For L_3: Z=(3000-6600)/2121.3= -1.697

$N(Z)$=.0945 (Table Lookup for Standard Normal)

$\int_{-\infty}^{Z} N(x)dx$ =.0448 (Table Lookup for Cumulative distribution of the Standard Normal)

The derivative terms are:

$N^{(2)}(-1.697)$= 0.1776
$N^{(3)}(-1.697)$= -0.0193
$N^{(5)}(-1.697)$= -0.8828

And the approximation to the Equivalent Load Duration Curve as in (B.24) at the loading point 3000 is:

$$ELDC_3(3000) = 1 - 0.0448 + \frac{-0.22}{3!}(0.1776)$$

$$- \frac{0.047}{4!}(-0.0193) - \frac{10(0.22)^2}{6!}(-.8828)$$

$$= .949$$

The second point is L_4 which is equal to 3800:

Z= -1.320

$N(Z)$= 0.1669

$$\int_{-\infty}^{Z} N(x)dx = .0934$$

$N^{(2)}(-1.320)$= 0.1239
$N^{(3)}(-1.320)$= -0.2771
$N^{(5)}(-1.320)$= 0.1348

$$ELDC_3(3800) = 1 - 0.0934 + \frac{-0.22}{3!}(0.1239)$$

$$- \frac{0.047}{4!}(-0.2771) - \frac{10(0.22)^2}{6!}(-.1348)$$

$$= .901$$

Therefore, expected generation of the fourth unit is

$$E\{g_{4t}\} = \tau\,(0.9)(800)(0.949+0.901)/2 = 666\ \tau \text{ MWh}$$

Expected energy purchased from outside resources is:

$$E(r) = 0.901(4000\text{-}3800)\ \tau = 180.2\ \tau \text{ MWh}$$

The dual multiplier π_{4t} is:

$$\pi_{4t} = \chi_t \tau \frac{(0.949 + 0.901)}{2} = 0.925\ \chi_t\ \tau \quad (\$/\text{MW})$$

The significant aspects of the cumulant method are given as follow:

1. The individual plant cumulants and load cumulants are calculated one time only and used repeatedly throughout the analysis.
2. The determination of the coefficients of Gram-Charlier approximation for the *ELDC* facing the i-th unit involves simple addition and manipulation of load and plant cumulants.
3. The solution of the Gram-Charlier approximation involves several table lookups and evaluation of three derivatives terms.
4. Evaluation of the *ELDC* at Z_{i-1} and Z_i allows for an estimate of the expected generation.

APPENDIX C

PHOTOVOLTAIC SYSTEM MODEL

C.1 Calculation of Radiation on Tilted Surface

The assessment of PV power in electric utilities begins with modeling the solar radiation resource. Here, the availability and collection of data on insulation, temperature, wind speed and cloud cover at a particular site is vital. Since, the long-term meteorological data are generally available in the form of radiation data on a horizontal surface, the corresponding calculation for a tilted surface will be needed. Various methods have attempted to model the solar irradiance. In this appendix, we compute a tilted surface radiation using the following steps:

Step 1. Calculate the global radiation outside the atmosphere on a horizontal surface (G_{oh}).

- The radiation outside the atmosphere on a horizontal surface G_{oh} is given by:

$$G_{oh} = E_o G_{sc} cos\theta_z \tag{C.1}$$

where

E_o excentricity factor $(r_o/r)^2$, with mean distance between sun and earth r_o ($1.496 \mathrm{x} 10^{11}$ m) and time dependent distance between sun and earth r

G_{sc} solar constant (1353 W/m^2)

θ_z zenith angle calculated using the Hollands/Maier model [May88]

The excentricity factor is calculated by Spancer's Fourier series [Spe65]:

$$E_o = 1.00010 + 0.034221 \cos\Gamma + 0.001280 \sin\Gamma + 0.000719 \cos 2\Gamma + 0.000077 \sin 2\Gamma \tag{C.2}$$

where Γ is the day angle (2π (day-1)/365), with the day $\in [1,365]$

Step 2. Calculate the diffuse fraction of the global radiation according to the Orgill/Holands [Org77] using the available radiation data on a horizontal plane (G_h) and G_{oh} from Step 1.

- The diffuse fraction of the global radiation according to the Orgill/Hollands [Org77] is:

$$\frac{G_{dh}}{G_h} = \begin{cases} 1.0 - 0.249k_T & \text{if } 0 < k_T < 0.35 \\ 1.557 - 1.84k_T & \text{if } 0.35 < k_T < 0.75 \\ 0.177 & \text{if } k_T > 0.75 \end{cases} \tag{C.3}$$

where k_T is clearness index defined as:

$$k_T = \frac{G_h}{G_{oh}}$$

G_{dh} diffuse radiation on a horizontal plane
G_{bh} beam radiation on a horizontal plane

Step 3. Calculate the position of the sun, such as azimuth, elevation, declination and hour angle.

Step 4. Using the output from Steps 2 and 3, calculate the radiation on a tilted surface (G_t) from the available radiation data on a horizontal plane. The solar radiation is time dependent as fluctuations in solar irradiance may occur in a short time.

- The global radiation on a tilted surface is calculated by the isotropic Liu/Jordan-model [Liu63] as:

$$G_t = G_{bh} \frac{\cos\theta}{\cos\theta_z} + \frac{1}{2} G_{dh}(1+\cos\beta) + \frac{1}{2}\rho G_h(1-\cos\beta) \tag{C.4}$$

where,
θ incidence angle of beam radiation
θ_z zenith angle
β slope of the plane
ρ ground reflectance

The first term of (C.4) represents beam radiation on tilted surface, the second is the diffuse radiation and the third term is the reflected radiation.

C.2 Photovoltaic Cell Model

We discuss the PV cell model that is used in conjunction with tilted surface radiation data. The calculation of the output current of a PV generator is dependent on the PV generator voltage, radiation on the PV generator, wind speed and ambient temperature. The relationship between voltage V_c of a cell of PV and current I_c is given by a diode model as:

$$I_c = I_L - I_o\left(\exp\left(\frac{q(V_c + I_c R_s)}{A k_b T_c}\right) - 1\right) - \frac{(V_c - I_c R_s)}{R_{sh}} \tag{C.5}$$

where
R_s series resistance parameter of the cell [Ωm^2]
R_{sh} shunt resistance parameter [Ωm^2]
A diode parameter
T_c cell temperature [^oK]
q charge of an electron ($1.6021 x 110^{-19}$ As)
k_b Boltzman Constant ($1.3854x10-23\ JK^{-1}$)

Equation (C.5) is applicable to a particular radiation and cell temperature. The light-generated current (I_L) is proportional to the global radiation (G_t) on the generator plane and is assumed to be linearly dependent on the cell temperature T_c:

$$I_L = (C_o + C_1 T_c)\, G_t \tag{C.6}$$

where

C_o coefficient of light-generated current density [V^{-1}]

C_1 temperature coefficient of light-generated current density [$V^{-1}K^{-1}$]

The dependence of the saturation current on temperature is given by:

$$I_o = C_{01} T_c^{\,3} exp\left(-\frac{1.12\, q}{k_b T_c} \right) \tag{C.7}$$

where C_{01} is the coefficient of saturation current density [$Am^{-2}K^{-3}$].

For a PV generator with N_p parallel cells and N_s cells connected in series, the resulting voltage and current is calculated as follows:

$$V_{pv} = N_s\, V_c \tag{C.8}$$

$$I_{pv} = N_p\, I_c \tag{C.9}$$

The PV current as a function of voltage for different radiation at 25°C is depicted in Figure C.1. This system has A=1; R_{sh}=0.130 Ωm^2; R_s=0.381Ωm^2; C_o=0.306 V^{-1}; C_1=0.179x10^{-4} $V^{-1}K^{-1}$; C_{01}=1.708x 10^{-4} $Am^{-2}K^{-3}$; Ac=0.01m^2; N_s=36; N_p=1.

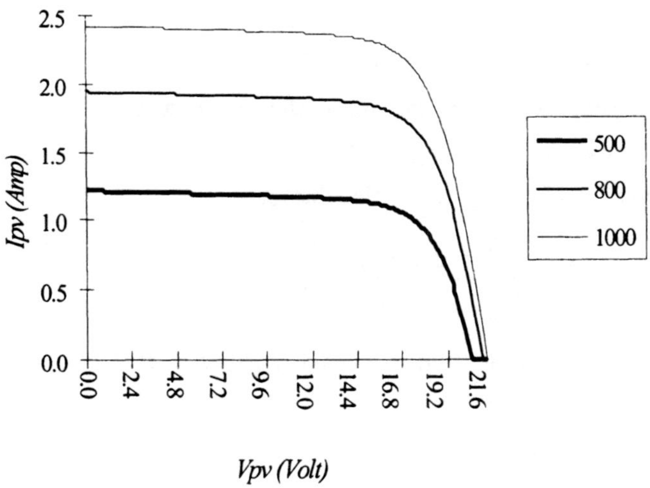

Figure C.1 Current vs. Voltage Characteristic for a PV Generator for Different W/m^2 Radiation at 25°C

The PV output is also affected by PV cell temperature. Given the radiation and wind speed data, we have to determine the PV cell temperature. The thermal model of PV is based on an energy balance.

$$m_{mod} N_{mod} c_{mod} \frac{dT_c}{dt} + P_{pv} = \dot{Q}_G - \dot{Q}_r - \dot{Q}_c \tag{C.10}$$

where

m_{mod} mass of a PV module [kg]

Nmod	number of cells in a PV module
c_{mod}	specific heat of a PV module [$Jkg^{-1}K^{-1}$]
P_{pv}	electrical power output of the generator [W]
$\dot{Q}_G$	insolation [W] on the whole generator
$\dot{Q}_r$	losses through radiation [W]
$\dot{Q}_c$	losses through convection [W]

The absorbed insolation is modeled by:

$$\dot{Q}_G = a\, G_t(t)\, A_{mod} N_{mod} \tag{C.11}$$

The coefficient a of absorption is assumed to be constant. If each PV module has area A_{mod} losses through radiation are modeled by:

$$\dot{Q}_r = 2\varepsilon A_{mod} N_{mod} \sigma (T_c^4 - T_{amb}^4) \tag{C.12}$$

where

ε	emission factor
σ	Stefan-Boltzman Constant ($5.6697 \times 10^{-8}\ Wm^{-2}K^{-4}$)

Losses through convection are given by:

$$\dot{Q}_c = 2\gamma\ A_{mod} N_{mod} (T_c - T_{amb}) \tag{C.13}$$

In case of free convection the heat loss coefficient γ is set to:

$$\gamma_f = 1.78 (T_c - T_{amb})^{1/3} \tag{C.14}$$

forced convection along l_{mod} is modeled through:

$$\gamma_w = \frac{4.77 \upsilon_w^{0.8} (l_{mod} N_{mod})^{-0.2}}{1 - 0.17 \upsilon_w^{-0.1} (l_{mod} N_{mod})^{-0.1}} \tag{C.15}$$

with wind speed v_w. In case of mixed convection γ is set to:

$$\gamma = \sqrt[3]{\gamma_f^3 + \gamma_w^3} \tag{C.16}$$

C.3 Battery Model

Usually a battery is treated as constant voltage storage. The energy produced by PV is computed first, and load is supplied with the excess energy available to charge the battery. If the system load exceeds the PV output, the difference is drawn from the battery until the battery is fully discharged. This charge accounting scenario may be repeated on an hourly basis. This procedure provides a reasonable approach, however, voltage variations in a battery are required for determining the operating point of the system. Correspondingly, we devise the following method to simulate the battery operation more accurately. The relationship between voltage V_b and current I_b for discharging ($I_b < 0$) the battery is:

$$V_b = V_{od} - g_d H + \rho_d \frac{I_b}{C_n}\left(1 + M_d \frac{H}{C_d - H}\right) \tag{C.17}$$

where

V_{od}	open circuit voltage (discharge) [V]
g_d	electrolyte coefficient (discharge) [V]
ρ_d	inner resistance parameter (discharge) [Ω Ah]
M_d	battery type coefficient (discharge)
C_d	capacity coefficient (discharge)
H	normalized depth of discharge $1-C/C_n$

For charging the battery (I_b>0), we use the following equation:

$$V_b = V_{oc} - g_c H + \rho_c \frac{I_b}{C_n}\left(1 + M_c \frac{F}{C_c - F}\right) \tag{C.18}$$

where

V_{oc}	open circuit voltage (charge) [V]
g_c	electrolyte coefficient (charge) [V]
ρ_c	inner resistance parameter (charge) [Ω Ah]
M_c	battery type coefficient (charge)
C_c	capacity coefficient (charge)
F	normalized state of charge C/C_n

In the case of charging, an ampere hour efficiency η_b is utilized. First, a current at which the gassing effect occurs is defined as:

$$I_m = \begin{cases} b_w C_n (1 - F) & \text{if } F > F_l \\ b_w C_n (1 - F_l) & \text{else} \end{cases} \tag{C.19}$$

where

b_w	Wood Parameter
F_l	limiting state of charge [V]
C_n	nominal capacity [Ah]

Then the charging efficiency is given as:

$$\eta_b = \begin{cases} a_w & \text{if } I_b \le I_m \\ \dfrac{a_w I_m}{I_b} & \text{else} \end{cases} \tag{C.20}$$

(η_b is set to 1 if battery is discharged). The battery capacity is updated with the charge/discharge current and the self-discharge current as follows:

$$C(t+\Delta t) = C(t) + \left(\eta_b(t)\, I_b - \frac{q_s\, C(t)}{100x24h} \right) \frac{\Delta t}{3600s} \qquad \text{(C.21)}$$

where a_w is the Wood parameter.

Figure C.2 shows a sample of battery characteristics for different values of the state of charge $F_N=C/C_n$. The following are the battery parameters: N_s=3; C_n=100 Ah; V_{oc}=2.089V; g_c=0.089V; ρ_c=0.999ΩAh; M_c=0.422; C_c=1.050; V_{od}=2.077V; g_d=0.0833V; ρ_d=0.374ΩAh; M_d=1.694; C_d=1.600; a_w=0.96; b_w=1.33h^{-1}; F_l=0.99925; q_s=0.166

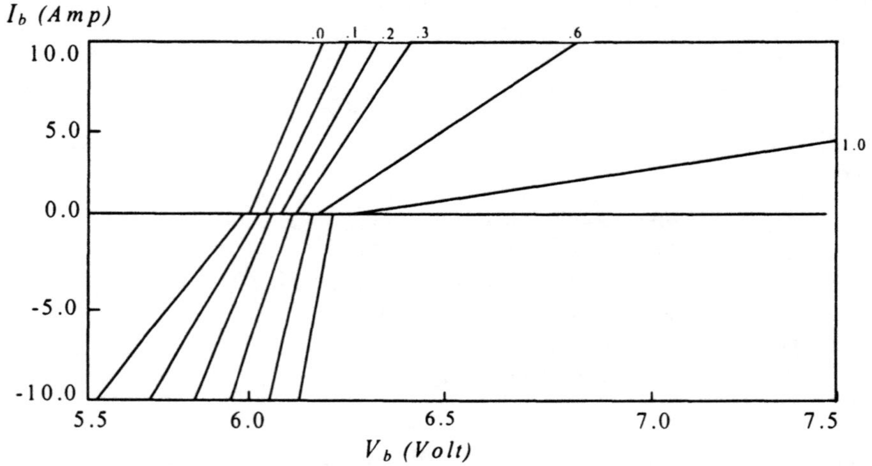

Figure C.2 Battery Characteristics. $F_N=C/C_n$ are 0.0, 0.2,... , 1.0

In above formulation the SOC (F_N) is assumed to be restricted between 0% and 100% . This is somewhat idealistic, since the battery's maximum capacity (C_n) does change with age, temperature, plate sulfation, electrolyte quality, etc. Since most battery expect a decreases in capacity with time, we may wish to let C_n decrease slightly in each day over the period that the model is run, as follows:

$$C_n = D_b\ C_{max} \qquad \text{(C.22)}$$

Lead acid battery capacity daily decline rates D_b are highly variable and have observed between 0.15%/day and 0.02%/day.

APPENDIX D

MONTE-CARLO SAMPLING ALGORITHM WITH GENERALIZED REGRESSION

Let the state of power system be represented by a vector $x=(x_1, x_2, ..., x_M)$, where x_i is the state of the i-th component. The set of M components typically includes the state of each system element (generators, lines and transformers) and the load level. X, the state space, denotes the set of all possible states x.

For every state x, it is possible to calculate its associated probability of occurrence, $P(x)$. For example, if component failures are statically independent, $P(x)$ is the product of the probabilities associated with the state of each component.

Finally, let $F(x)$ represent a test applied to a given configuration x. The objective of the test is to determine the energy reserve at that specific configuration of generators and lines while supplying the demand. Because system states are random variables, the test result is also a random variable. The expected value and the variance of the test result are given by:

$$E(F) = \sum_{x \in X} F(x) P(x) \tag{D.1}$$

$$V(F) = \sum_{x \in X} (F(x) - E(F))^2 P(x) \tag{D.2}$$

In the Monte-Carlo method, state x is sampled from the joint probability distribution. The expected value of F, $E(F)$, is estimated as:

$$\hat{E}(F) = \frac{1}{NS} \sum_{i=1}^{NS} F(x^i) \tag{D.3}$$

where:
- $\hat{E}(F)$ is the estimate of the expected value
- NS is the number of samplings
- $F(x_i)$ is the result of the test function applied to the i-th sampled state
- $F(x_i)$ is the result of the test function applied to the i-th sampled state

It is important to observe that $\hat{E}(F)$ is not the "true" (population) expected value $E(F)$ (which is usually unknown) but an estimate of this value. Because x and $F(x)$ are random variables, the estimator $\hat{E}(F)$ which is an average of NS samples of $F(x)$ is also a random variable. The expected value and variance of $\hat{E}(F)$ are:

$$E(\hat{E}(F)) = E(F) \tag{D.4}$$

$$V(\hat{E}(F)) = V(F) / NS \tag{D.5}$$

Equation (D.4) indicates that the estimator $\hat{E}(F)$ is unbiased, that is, its expected value is equal to the population's expected value $E(F)$. Equation (D.5) indicates that the variance of the estimate which is a measure of its uncertainty, depends on the variance of random variable, $V(F)$. Equation (D.5) also indicates that the uncertainty of the estimate decreases as the number of samplings NS increases.

The uncertainty around the estimate is often represented as a relative uncertainty, or coefficient of variation:

$$\beta = (V(\hat{E}(F)))^{0.5} / \hat{E}(F) \tag{D.6}$$

Substituting (D.5) into (D.6), we obtain:

$$\beta = (V(F) / NS)^{0.5} / \hat{E}(F) \tag{D.7}$$

Rewriting (D.7) in term of NS results in:

$$NS = V(F) / (\beta \hat{E}(F))^2 \tag{D.8}$$

Equation (D.8) indicates that the number of samplings NS depends on the variance of the random variable and on the required accuracy, β. If we wish to reduce the number of samplings while keeping the same accuracy, we must find ways to reduce $V(F)$. Let Z be a random variable, which is strongly correlated with F. Define a new random variable Y as:

$$Y = F - Z + E(Z) \tag{D.9}$$

It is easy to see that Y and F have the same expected value:

$$E(Y) = E(F) - E(Z) + E(Z) = E(F) \tag{D.10}$$

$$V(Y) = V(F) + V(Z) - 2\,C(F,Z) \tag{D.11}$$

where $C(F,Z)$ indicates the covariance between F and Z.

If F and Z are strongly correlated, the negative covariance factor 2 $C(F,Z)$ in (D.11) may outweigh the positive factor $V(Z)$. As a consequence, $V(Y)$ may be smaller than $V(F)$ and $V(Y)$ may be smaller than $V(F)$.

Because Y and F have the same expected values, $E(F)$ may then be more efficiently estimated by sampling from Y:

$$\hat{E}(Y) = \frac{1}{NS} \sum_{i=1}^{NS} Y^i \tag{D.12}$$

As seen in (D.8), the reduction in number of samplings is proportional to the variances:

$$NS'/NS = V(Y)/V(F) \tag{D.13}$$

APPENDIX E

IEEE 118-BUS TEST SYSTEM DATA

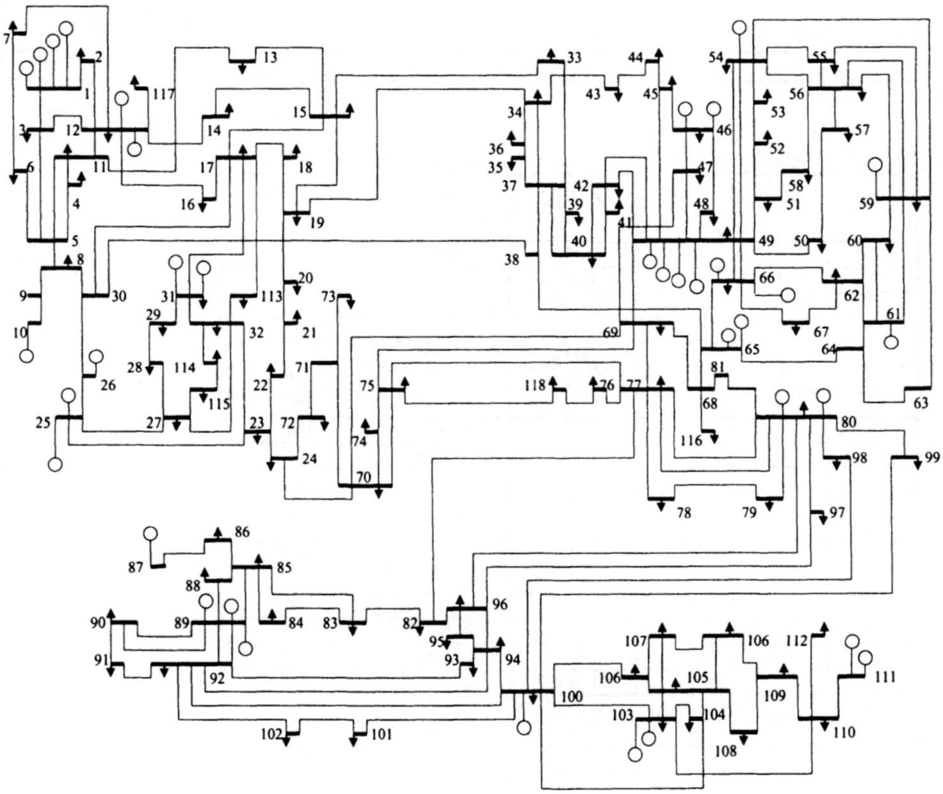

Figure E.1 IEEE118-Bus System

Table E.1 Weekly Peak Load in Percent of Annual Peak

Week	Peak Load	Week	Peak Load
18	83.7	24	88.7
19	87.0	25	89.6
20	88.0	26	86.1
21	85.6	27	75.5
22	81.1	28	81.6
23	90.0	29	80.1

Table E.2 Daily Transactions in week 27

τ	*P* (MW)	*Q* (MVAR)	τ	*P* (MW)	*Q* (MVAR)	τ	*P* (MW)	*Q* (MVAR)
1	4242	2969.4	9	4950	3465.0	17	4351	3045.7
2	3916	2741.2	10	5438	3806.6	18	4786	3350.2
3	3698	2588.6	11	5385	3769.5	19	4895	3426.5
4	3589	2512.3	12	5276	3693.2	20	4950	3465.0
5	3481	2436.7	13	5112	3578.4	21	4895	3426.5
6	3484	2438.8	14	5003	3502.1	22	4786	3350.2
7	3589	2512.3	15	4732	3312.4	23	4732	3312.4
8	3807	2664.9	16	4406	3084.2	24	4732	3312.4

Table E.3 Parameters of the Transmission Lines in Maintenance Area*

Line	Bus	Bus	R(pu)	X (pu)	Length (km)
1	3	5	0.0241	0.1080	150
2	5	6	0.0119	0.0540	70
3	6	7	0.0046	0.0208	50
4	8	9	0.0024	0.0305	580
5	3	12	0.0484	0.1600	200
6	9	10	0.0026	0.0322	600
7	4	11	0.0209	0.0688	100
8	5	11	0.0203	0.0682	100
9	11	12	0.0060	0.0196	50
10	2	12	0.0187	0.0616	80
11	4	5	0.0018	0.0080	100
12	8	5	0.0000	0.0267	100
13	7	12	0.0086	0.0340	100
14	11	13	0.0223	0.0731	200
15	12	14	0.0215	0.0707	200
16	13	15	0.0744	0.2444	600
17	14	15	0.0595	0.1950	200
18	12	16	0.0212	0.0834	100
19	15	17	0.0132	0.0437	100
20	16	17	0.0454	0.1801	100
21	17	18	0.0123	0.0505	100
22	18	19	0.0112	0.0493	100
23	19	20	0.0252	0.1170	100
24	15	19	0.0120	0.0394	100

* IEEE-118 Bus System Data is given in [Was96]

Table E.4 Line Maintenance Cost in week 27

τ	($/km)	τ	($/km)	τ	($/km)
1	3.5	9	1.152	17	1.152
2	3.5	10	1.152	18	2.3
3	3.5	11	1.152	19	2.3
4	3.5	12	1.152	20	2.3
5	3.5	13	1.152	21	2.3
6	3.5	14	1.152	22	2.3
7	2.3	15	1.152	23	3.5
8	2.3	16	1.152	24	3.5

Table E.5 Recallable Contract in week 27

Bus	+j (%)*	($/MWh)	($/MVAR)
15	2.12	11.87	1.18
42	2.26	8.40	0.84
49	2.05	16.21	1.62
54	2.66	18.47	1.84
59	6.53	8.98	0.90
74	1.6	27.88	2.79
80	3.06	21.40	2.14
89	3.84	17.96	1.80
112	1.60	19.28	1.93
116	4.34	9.53	0.95

*) percent of total hourly load in Table E.2

Table E.6 Ancillary Services for Reactive Power in week 27

Bus	(MVAR)	($/MVAR)
65	277	1.3
66	277	1.2
69	465	1.1
80	390	1.0
89	550	0.8
100	230	0.9

Table E.7 Characteristics of Generating Units Part (I)

No							
1	12.	2.40	0.02533	25.5472	24.3891	1.0	0.
2	20.	4.00	0.01561	37.9637	118.9083	0.0	30.
3	20.	4.00	0.01359	37.7770	118.4576	0.0	30.
4	20.	4.00	0.01161	37.9637	118.9083	0.0	30.
5	20.	4.00	0.01059	38.7770	119.4576	0.0	30.
6	20.	4.00	0.01199	37.5510	117.7551	0.0	30.
7	20.	4.00	0.01261	37.6637	118.1083	0.0	30.
8	76.	15.20	0.00962	13.5073	81.8259	0.0	80.
9	76.	15.20	0.00876	13.3272	81.1364	1.0	80.
10	76.	15.20	0.00895	13.3538	81.2980	0.0	80.
11	76.	15.20	0.00932	13.4073	81.6259	0.0	80.
12	100.	25.00	0.00623	18.0000	217.8952	0.0	100.
13	100.	25.00	0.00599	18.6000	219.7752	0.0	100.
14	100.	25.00	0.00612	18.1000	218.3350	0.0	100.
15	100.	25.00	0.00588	18.2800	216.7752	0.0	100.
16	100.	25.00	0.00598	18.2000	218.7752	0.0	100.
17	100.	25.00	0.00578	17.2800	216.7752	1.0	100.
18	100.	25.00	0.00698	19.2000	218.7752	0.0	100.
19	155.	54.25	0.00473	10.7154	143.0288	0.0	200.
20	155.	54.25	0.00481	10.7367	143.3179	0.0	200.
21	155.	54.25	0.00487	10.7583	143.5972	0.0	200.
22	197.	68.95	0.00259	23.0000	259.1310	1.0	300.
23	197.	68.95	0.00260	23.1000	259.6490	1.0	300.
24	197.	68.95	0.00263	23.2000	260.1760	0.0	300.
25	197.	68.95	0.00264	23.4000	260.5760	0.0	300.
26	197.	68.95	0.00267	23.5000	261.1760	0.0	300.
27	197.	68.95	0.00261	23.0400	260.0760	0.0	300.
28	350.	140.00	0.00150	10.8416	176.0575	0.0	500.
29	350.	140.00	0.00153	10.8616	177.0575	0.0	500.
30	350.	140.00	0.00143	10.6616	176.0575	0.0	500.
31	350.	140.00	0.00163	10.9616	177.9575	1.0	500.
32	400.	100.00	0.00194	7.4921	310.0021	0.0	800.
33	400.	100.00	0.00195	7.5031	311.9102	0.0	800.
34	400.	100.00	0.00196	7.5121	312.9102	0.0	800.
35	400.	100.00	0.00197	7.5321	314.9102	0.0	800.
36	400.	100.00	0.00199	7.6121	313.9102	0.0	800.

Where,

fuel cost of fuel type one ($/MBtu)

start up cost ($)

Table E.8 Characteristics of Generating Units Part (II)

No	Up	Dn	Ini	UR	DR	Bus	FL	FU		
1	1	-1	-1	12.0	12.0	31	0.0	19999	0.00	1.15
2	1	-1	-1	20.0	20.0	46	0.0	999999	0.00	1.1
3	1	-1	-1	20.0	20.0	46	0.0	999999	0.00	1.1
4	1	-1	-1	20.0	20.0	103	0.0	999999	0.00	1.1
5	1	-1	-1	20.0	20.0	87	0.0	999999	0.00	1.1
6	1	-1	-1	20.0	20.0	31	0.0	19999	0.00	1.12
7	1	-1	-1	20.0	20.0	111	0.0	999999	0.00	1.1
8	3	-2	3	38.0	38.0	12	0.0	999999	0.35	1.1
9	3	-2	3	38.0	38.0	12	0.0	19999	0.35	1.12
10	3	-2	3	38.0	38.0	1	0.0	999999	0.35	1.1
11	3	-2	3	38.0	38.0	54	0.0	999999	0.35	1.1
12	4	-2	5	50.0	50.0	25	0.0	999999	0.00	1.1
13	4	-2	5	50.0	50.0	26	0.0	999999	0.00	1.1
14	4	-2	5	50.0	50.0	65	0.0	19999	0.00	1.25
15	4	-2	-3	50.0	50.0	66	0.0	999999	0.30	1.1
16	4	-2	-3	50.0	50.0	100	0.0	999999	0.30	1.1
17	4	-2	-3	50.0	50.0	111	0.0	19999	0.30	1.25
18	4	-2	-3	50.0	50.0	103	0.0	19999	0.30	1.
19	5	-3	5	77.5	77.5	10	0.0	21999	0.00	1.
20	5	-3	5	77.5	77.5	49	0.0	999999	0.00	1.1
21	5	-3	5	77.5	77.5	49	0.0	999999	0.25	1.1
22	5	-4	-4	98.5	98.5	25	0.0	19999	0.25	1.27
23	5	-4	-4	98.5	98.5	59	0.0	19999	0.25	1.21
24	5	-4	-4	98.5	98.5	61	0.0	999999	0.25	1.1
25	5	-4	-4	98.5	98.5	80	0.0	999999	0.00	1.1
26	5	-4	-4	98.5	98.5	89	0.0	999999	0.00	1.1
27	5	-4	-4	98.5	98.5	1	0.0	999999	0.00	1.1
28	8	-5	10	175.0	175.0	1	0.0	999999	0.20	1.1
29	8	-5	10	175.0	175.0	49	0.0	999999	0.20	1.1
30	8	-5	10	175.0	175.0	49	0.0	999999	0.20	1.1
31	8	-5	10	175.0	175.0	89	0.0	999999	0.20	1.29
32	8	-5	10	200.0	200.0	65	0.0	999999	0.15	1.1
33	8	-5	10	200.0	200.0	66	0.0	999999	0.15	1.1
34	8	-5	10	200.0	200.0	1	0.0	39999	0.15	1.39
35	8	-5	10	200.0	200.0	80	0.0	999999	0.15	1.1
36	8	-5	10	200.0	200.0	89	0.0	999999	0.15	1.1

Where,

Up Minimum up time for unit *i* (hours)

Dn Minimum down time for unit *i* (hours)

Ini Initial condition (hours)

UR Ramp up rate (MW/hr)

DR Ramp down rate (MW/hr)

Bus Bus where the unit is located in the system

FL Lower limit for fuel constraint (MBtu)

FU Upper limit for fuel constraint (MBtu)

Coefficient for emission function (lbs/MBtu)

where

fuel cost of fuel type two ($/MBtu)

Table E.9 24-Hours Load and Spinning Reserve Distribution

Hr	P_D	Rs	hr	P_D	Rs	hr	P_D	Rs
1	3309	165.4	9	3394	169.7	17	3860	193.0
2	3054	152.7	10	3733	186.6	18	4242	212.1
3	2885	144.2	11	3818	190.9	19	4200	210.0
4	2800	140.0	12	3860	193.0	20	4115	205.7
5	2715	135.7	13	3818	190.9	21	3987	199.4
6	2757	137.9	14	3733	186.6	22	3903	195.1
7	2800	140.0	15	3691	184.5	23	3691	184.5
8	2969	148.5	16	3691	184.5	24	3436	171.8

Where,
P_D: Load at hour (hr)
Rs: Reserve at hour (hr)

Table E.10 AC Load Flow Data

No.	V	Angle	Pg	Qg	Pd	Qd	Type
1	1.035	30	5.164	0	0	0	2
2	0.971	11.22	0	0	0.2	0.09	0
3	0.968	11.56	0	0	0.39	0.1	0
4	0.998	15.28	0	0	0.39	0.12	0
5	1.002	15.73	0	0	0	0	0
6	0.99	13	0	0	0.52	0.22	0
7	0.989	12.56	0	0	0.19	0.02	0
8	1.015	20.77	0	0	0.28	0	0
9	1.043	28.02	0	0	0	0	0
10	1.05	35.61	4.5	0	0	0	1
11	0.985	12.72	0	0	0.7	0.23	0
12	0.99	12.2	0.85	0	0.47	0.1	1
13	0.968	11.35	0	0	0.34	0.16	0
14	0.984	11.5	0	0	0.14	0.01	0
15	0.97	11.23	0	0	0.9	0.3	0
16	0.984	11.91	0	0	0.25	0.1	0
17	0.995	13.74	0	0	0.11	0.03	0
18	0.973	11.53	0	0	0.6	0.34	0
19	0.963	11.05	0	0	0.45	0.25	0
20	0.958	11.93	0	0	0.18	0.03	0
21	0.959	13.52	0	0	0.14	0.08	0
22	0.97	16.08	0	0	0.1	0.05	0
23	1	21	0	0	0.07	0.03	0
24	0.992	20.89	0	0	0.13	0	0
25	1.05	27.93	2.2	0	0	0	1
26	1.015	29.71	3.14	0	0	0	1
27	0.968	15.35	0	0	0.71	0.13	0
28	0.962	13.62	0	0	0.17	0.07	0
29	0.963	12.63	0	0	0.24	0.04	0
30	0.968	18.79	0	0	0	0	0
31	0.967	12.75	0.07	0	0.43	0.27	1
32	0.964	14.8	0	0	0.59	0.23	0
33	0.972	10.63	0	0	0.23	0.09	0
34	0.986	11.3	0	0	0.59	0.26	0
35	0.981	10.87	0	0	0.33	0.09	0
36	0.98	10.87	0	0	0.31	0.17	0
37	0.992	11.77	0	0	0	0	0
38	0.962	16.91	0	0	0	0	0
39	0.97	8.41	0	0	0.27	0.11	0
40	0.97	7.35	0	0	0.66	0.23	0

41	0.967	6.92	0	0	0.37	0.1	0
42	0.985	8.53	0	0	0.96	0.23	0
43	0.978	11.28	0	0	0.18	0.07	0
44	0.985	13.82	0	0	0.16	0.08	0
45	0.987	15.67	0	0	0.53	0.22	0
46	1.005	18.49	0.19	0	0.28	0.1	1
47	1.017	20.73	0	0	0.34	0	0
48	1.021	19.93	0	0	0.2	0.11	0
49	1.025	20.94	2.04	0	0.87	0.3	1
50	1.001	18.9	0	0	0.17	0.04	0
51	0.967	16.28	0	0	0.17	0.08	0
52	0.957	15.32	0	0	0.18	0.05	0
53	0.946	14.35	0	0	0.23	0.11	0
54	0.955	15.26	0.48	0	1.13	0.32	1
55	0.952	14.97	0	0	0.63	0.22	0
56	0.954	15.16	0	0	0.84	0.18	0
57	0.971	16.36	0	0	0.12	0.03	0
58	0.959	15.51	0	0	0.12	0.03	0
59	0.985	19.37	1.55	0	2.77	1.13	1
60	0.993	23.15	0	0	0.78	0.03	0
61	0.995	24.04	1.6	0	0	0	1
62	0.998	23.43	0	0	0.77	0.14	0
63	0.969	22.75	0	0	0	0	0
64	0.984	24.52	0	0	0	0	0
65	1.005	27.65	3.91	0	0	0	1
66	1.05	27.48	3.92	0	0.39	0.18	1
67	1.02	24.84	0	0	0.28	0.07	0
68	1.003	27.55	0	0	0	0	0
69	0.955	10.67	0	0	0.51	0.27	0
70	0.984	22.58	0	0	0.66	0.2	0
71	0.987	22.15	0	0	0	0	0
72	0.98	20.98	0	0	0.12	0	0
73	0.991	21.94	0	0	0.06	0	0
74	0.958	21.64	0	0	0.68	0.27	0
75	0.967	22.91	0	0	0.47	0.11	0
76	0.943	21.77	0	0	0.68	0.36	0
77	1.006	26.72	0	0	0.61	0.28	0
78	1.003	26.42	0	0	0.71	0.26	0
79	1.009	26.72	0	0	0.39	0.32	0
80	1.04	28.96	4.77	0	1.3	0.26	1
81	0.997	28.1	0	0	0	0	0
82	0.989	27.24	0	0	0.54	0.27	0
83	0.985	28.42	0	0	0.2	0.1	0
84	0.98	30.95	0	0	0.11	0.07	0
85	0.985	32.51	0	0	0.24	0.15	0
86	0.987	31.14	0	0	0.21	0.1	0
87	1.015	31.4	0.04	0	0	0	1
88	0.987	35.64	0	0	0.48	0.1	0
89	1.005	39.69	6.07	0	0	0	1
90	0.985	33.29	0	0	1.63	0.42	0
91	0.98	33.31	0	0	0.1	0	0
92	0.993	33.8	0	0	0.65	0.1	0
93	0.987	30.79	0	0	0.12	0.07	0
94	0.991	28.64	0	0	0.3	0.16	0
95	0.981	27.67	0	0	0.42	0.31	0
96	0.993	27.51	0	0	0.38	0.15	0
97	1.011	27.88	0	0	0.15	0.09	0
98	1.024	27.4	0	0	0.34	0.08	0
99	1.01	27.04	0	0	0.42	0	0
100	1.017	28.03	2.52	0	0.37	0.18	1
101	0.993	29.61	0	0	0.22	0.15	0
102	0.991	32.3	0	0	0.05	0.03	0
103	1.001	24.44	0.4	0	0.23	0.16	1

104	0.971	21.69	0	0	0.38	0.25	0
105	0.965	20.57	0	0	0.31	0.26	0
106	0.962	20.32	0	0	0.43	0.16	0
107	0.952	17.53	0	0	0.5	0.12	0
108	0.967	19.38	0	0	0.02	0.01	0
109	0.967	18.93	0	0	0.08	0.03	0
110	0.973	18.09	0	0	0.39	0.3	0
111	0.98	19.74	0.36	0	0	0	1
112	0.975	14.99	0	0	0.68	0.13	0
113	0.993	13.74	0	0	0.06	0	0
114	0.96	14.46	0	0	0.08	0.03	0
115	0.96	14.46	0	0	0.22	0.07	0
116	1.005	27.12	0	0	1.84	0	0
117	0.974	10.67	0	0	0.2	0.08	0
118	0.949	21.92	0	0	0.33	0.15	0

Where,
Type of bus: 0=PQ, 1=PV, 2=Slack

Table E.11 Network Data Part (I)

From	To	R	X	Bch/2	Tap	Rating
1	2	0.0303	0.0999	0.0127	0	3.0
1	3	0.0129	0.0424	0.0054	0	3.0
4	5	0.0018	0.0080	0.0010	0	3.0
3	5	0.0241	0.1080	0.0142	0	3.0
5	6	0.0119	0.0540	0.0071	0	3.0
6	7	0.0046	0.0208	0.0027	0	3.0
8	9	0.0024	0.0305	0.5810	0	3.0
8	5	0.0000	0.0267	0.0000	0.985	3.0
9	10	0.0026	0.0322	0.6150	0	3.0
4	11	0.0209	0.0688	0.0087	0	3.0
5	11	0.0203	0.0682	0.0087	0	3.0
11	12	0.0060	0.0196	0.0025	0	3.0
2	12	0.0187	0.0616	0.0079	0	3.0
3	12	0.0484	0.1600	0.0203	0	3.0
7	12	0.0086	0.0340	0.0044	0	3.0
11	13	0.0223	0.0731	0.0094	0	3.0
12	14	0.0215	0.0707	0.0091	0	3.0
13	15	0.0744	0.2444	0.0313	0	3.0
14	15	0.0595	0.1950	0.0251	0	3.0
12	16	0.0212	0.0834	0.0107	0	3.0
15	17	0.0132	0.0437	0.0222	0	3.0
16	17	0.0454	0.1801	0.0233	0	3.0
17	18	0.0123	0.0505	0.0065	0	3.0
18	19	0.0112	0.0493	0.0057	0	3.0
19	20	0.0252	0.1170	0.0149	0	3.0
15	19	0.0120	0.0394	0.0050	0	3.0
20	21	0.0183	0.0849	0.0108	0	3.0
21	22	0.0209	0.0970	0.0123	0	3.0
22	23	0.0342	0.1590	0.0202	0	3.0
23	24	0.0135	0.0492	0.0249	0	3.0
23	25	0.0156	0.0800	0.0432	0	3.0
26	25	0.0000	0.0382	0.0000	0.96	3.0
25	27	0.0318	0.1630	0.0882	0	3.0
27	28	0.0191	0.0855	0.0108	0	3.0
28	29	0.0237	0.0943	0.0119	0	3.0
30	17	0.0000	0.0388	0.0000	0.96	3.0

8	30	0.0043	0.0504	0.2570	0	3.0
17	31	0.0474	0.1563	0.0200	0	3.0
29	31	0.0108	0.0331	0.0041	0	3.0
23	32	0.0317	0.1153	0.0586	0	3.0
31	32	0.0298	0.0985	0.0126	0	3.0
27	32	0.0229	0.0755	0.0096	0	3.0
15	33	0.0380	0.1244	0.0160	0	3.0
19	34	0.0752	0.2470	0.0316	0	3.0
35	36	0.0022	0.0102	0.0013	0	3.0
35	37	0.0110	0.0497	0.0066	0	3.0
33	37	0.0415	0.1420	0.0183	0	3.0
34	36	0.0087	0.0268	0.0028	0	3.0
34	37	0.0026	0.0094	0.0049	0	3.0
37	39	0.0321	0.1060	0.0135	0	3.0
37	40	0.0593	0.1680	0.0210	0	3.0
30	38	0.0046	0.0540	0.2110	0	3.0
39	40	0.0184	0.0605	0.0078	0	3.0
40	41	0.0145	0.0487	0.0061	0	3.0
40	42	0.0555	0.1830	0.0233	0	3.0
41	42	0.0410	0.1350	0.0172	0	3.0
43	44	0.0608	0.2454	0.0303	0	3.0
34	43	0.0413	0.1681	0.0211	0	3.0
44	45	0.0224	0.0901	0.0112	0	3.0
45	46	0.0400	0.1356	0.0166	0	3.0
46	47	0.0380	0.1270	0.0158	0	3.0
46	48	0.0601	0.1890	0.0236	0	3.0
47	49	0.0191	0.0625	0.0080	0	3.0
42	49	0.0715	0.3230	0.0430	0	3.0
42	49	0.0715	0.3230	0.0430	0	3.0
45	49	0.0684	0.1860	0.0222	0	3.0
48	49	0.0179	0.0505	0.0063	0	3.0
49	50	0.0267	0.0752	0.0094	0	3.0
49	51	0.0486	0.1370	0.0171	0	3.0
51	52	0.0203	0.0588	0.0070	0	3.0
52	53	0.0405	0.1635	0.0203	0	3.0
53	54	0.0263	0.1220	0.0155	0	3.0
49	54	0.0730	0.2890	0.0369	0	3.0
49	54	0.0869	0.2910	0.0365	0	3.0
54	55	0.0169	0.0707	0.0101	0	3.0
54	56	0.0027	0.0095	0.0037	0	3.0
55	56	0.0049	0.0151	0.0019	0	3.0
56	57	0.0343	0.0966	0.0121	0	3.0
50	57	0.0474	0.1340	0.0166	0	3.0
56	58	0.0343	0.0966	0.0121	0	3.0
51	58	0.0255	0.0719	0.0089	0	3.0
54	59	0.0503	0.2293	0.0299	0	3.0
56	59	0.0825	0.2510	0.0284	0	3.0
56	59	0.0803	0.2390	0.0268	0	3.0
55	59	0.0474	0.2158	0.0282	0	3.0
59	60	0.0317	0.1450	0.0188	0	3.0
59	61	0.0328	0.1500	0.0194	0	3.0
60	61	0.0026	0.0135	0.0073	0	3.0
60	62	0.0123	0.0561	0.0073	0	3.0
61	62	0.0082	0.0376	0.0049	0	3.0
63	59	0.0000	0.0386	0.0000	0.96	3.0
63	64	0.0017	0.0200	0.1080	0	3.0
64	61	0.0000	0.0268	0.0000	0.985	3.0
38	65	0.0090	0.0986	0.5230	0	3.0

64	65	0.0027	0.0302	0.1900	0	3.0
49	66	0.0180	0.0919	0.0124	0	3.0
49	66	0.0180	0.0919	0.0124	0	3.0
62	66	0.0482	0.2180	0.0289	0	3.0
62	67	0.0258	0.1170	0.0155	0	3.0
65	66	0.0000	0.0370	0.0000	0.935	3.0
66	67	0.0224	0.1015	0.0134	0	3.0
47	69	0.0844	0.2778	0.0355	0	3.0
49	69	0.0985	0.3240	0.0414	0	3.0
68	69	0.0000	0.0370	0.0000	0.935	3.0
69	70	0.0300	0.1270	0.0610	0	3.0
24	70	0.0022	0.4115	0.0510	0	3.0
70	71	0.0088	0.0355	0.0044	0	3.0
24	72	0.0488	0.1960	0.0244	0	3.0
71	72	0.0446	0.1800	0.0222	0	3.0
71	73	0.0087	0.0454	0.0059	0	3.0
70	74	0.0401	0.1323	0.0168	0	3.0
70	75	0.0428	0.1410	0.0180	0	3.0
69	75	0.0405	0.1220	0.0620	0	3.0
74	75	0.0123	0.0406	0.0052	0	3.0
76	77	0.0444	0.1480	0.0184	0	3.0
69	77	0.0309	0.1010	0.0519	0	3.0
75	77	0.0601	0.1999	0.0249	0	3.0
77	78	0.0038	0.0124	0.0063	0	3.0
78	79	0.0055	0.0244	0.0032	0	3.0
77	80	0.0170	0.0485	0.0236	0	3.0
77	80	0.0294	0.1050	0.0114	0	3.0
79	80	0.0156	0.0704	0.0093	0	3.0
68	81	0.0018	0.0202	0.4040	0	3.0
81	80	0.0000	0.0370	0.0000	0.935	3.0
77	82	0.0298	0.0853	0.0409	0	3.0
82	83	0.0112	0.0366	0.0190	0	3.0
83	84	0.0625	0.1320	0.0129	0	3.0
83	85	0.0430	0.1480	0.0174	0	3.0
84	85	0.0302	0.0641	0.0062	0	3.0
85	86	0.0350	0.1230	0.0138	0	3.0
86	87	0.0283	0.2074	0.0223	0	3.0
85	88	0.0200	0.1020	0.0138	0	3.0
85	89	0.0239	0.1730	0.0235	0	3.0
88	89	0.0139	0.0712	0.0097	0	3.0
89	90	0.0518	0.1880	0.0264	0	3.0
89	90	0.0238	0.0997	0.0530	0	3.0
90	91	0.0254	0.0836	0.0107	0	3.0
89	92	0.0099	0.0505	0.0274	0	3.0
89	92	0.0393	0.1581	0.0207	0	3.0
91	92	0.0387	0.1272	0.0163	0	3.0
92	93	0.0258	0.0848	0.0109	0	3.0
92	94	0.0481	0.1580	0.0203	0	3.0
93	94	0.0223	0.0732	0.0094	0	3.0
94	95	0.0132	0.0434	0.0055	0	3.0
80	96	0.0356	0.1820	0.0247	0	3.0
82	96	0.0162	0.0530	0.0272	0	3.0
94	96	0.0269	0.0869	0.0115	0	3.0
80	97	0.0183	0.0934	0.0127	0	3.0
80	98	0.0238	0.1080	0.0143	0	3.0
80	99	0.0454	0.2060	0.0273	0	3.0
92	100	0.0648	0.2950	0.0236	0	3.0
94	100	0.0178	0.0580	0.0302	0	3.0

95	96	0.0171	0.0547	0.0074	0	3.0
96	97	0.0173	0.0885	0.0120	0	3.0
98	100	0.0397	0.1790	0.0238	0	3.0
99	100	0.0180	0.0813	0.0108	0	3.0
100	101	0.0277	0.1262	0.0164	0	3.0
92	102	0.0123	0.0559	0.0073	0	3.0
101	102	0.0246	0.1120	0.0147	0	3.0
100	103	0.0160	0.0525	0.0268	0	3.0
100	104	0.0451	0.2040	0.0270	0	3.0
103	104	0.0466	0.1584	0.0203	0	3.0
103	105	0.0535	0.1625	0.0204	0	3.0
100	106	0.0605	0.2290	0.0310	0	3.0
104	105	0.0099	0.0378	0.0049	0	3.0
105	106	0.0140	0.0547	0.0072	0	3.0
105	107	0.0530	0.1830	0.0236	0	3.0
105	108	0.0261	0.0703	0.0092	0	3.0
106	107	0.0530	0.1830	0.0236	0	3.0
108	109	0.0105	0.0288	0.0038	0	3.0
103	110	0.0391	0.1813	0.0231	0	3.0
109	110	0.0278	0.0762	0.0101	0	3.0
110	111	0.0220	0.0755	0.0100	0	3.0
110	112	0.0247	0.0640	0.0310	0	3.0
17	113	0.0091	0.0301	0.0038	0	3.0
32	113	0.0615	0.2030	0.0259	0	3.0
32	114	0.0135	0.0612	0.0081	0	3.0
27	115	0.0164	0.0741	0.0099	0	3.0
114	115	0.0023	0.0104	0.0014	0	3.0
68	116	0.0003	0.0041	0.0820	0	3.0
12	117	0.0329	0.1400	0.0179	0	3.0
75	118	0.0145	0.0481	0.0060	0	3.0
76	118	0.0164	0.0544	0.0068	0	3.0
38	37	0.0000	0.0375	0.0000	0.935	3.0
26	30	0.0080	0.0860	0.4540	0	3.0
65	68	0.0014	0.0160	0.3190	0	3.0

Table E.12 Network Data Part (II)

Bus	G	jB
5	0	-0.4
34	0	0.14
37	0	-0.25
44	0	0.1
45	0	0.1
46	0	0.1
48	0	0.15
74	0	0.12
79	0	0.2
82	0	0.2
83	0	0.1
105	0	0.2
107	0	0.06
110	0	0.06

BIBLIOGRAPHY

[Abd 93] Abdul-Rahman, K.H., Shahidehpour, M., "A Fuzzy-Based Optimal Reactive Power Control," IEEE Transactions on Power Systems, Vol. 8, No. 2, pp. 662-670, May 1993.

[Abd 94a] Abdul-Rahman, K.H., Shahidehpour, M., "Application of Fuzzy Sets to Optimal Reactive Power Planning with Security Constraints," IEEE Transactions on Power Systems, Vol. 9, No. 2, pp. 589-597, May 1994.

[Abd 94b] Abdul-Rahman, K.H., Shahidehpour, M., "Reactive Power Optimization Using Fuzzy Load Representation," IEEE Transactions on Power Systems, Vol. 9, No. 2, pp. 898-905, May 1994.

[Abd 95a] Abdul-Rahman, K.H., Shahidehpour, M., and M. Daneshdoost, "AI Approach to Optimal Var Control with Fuzzy Reactive Loads," IEEE Transactions on Power Systems, Vol. 10, No. 1, pp. 88-97, Feb. 1995.

[Abd 95b] Abdul-Rahman, K.H., Shahidehpour, M., "Static Security in Power System Operation with Fuzzy Real Load Conditions," IEEE Transactions on Power Systems, Vol. 10, No. 1, pp. 77-87, Feb. 1995.

[Abd96] Abdul-Rahman, K.H., Shahidehpour, M., Agangic, M., and Mokhtari, S., "A Practical Resource Scheduling With OPF Constraints" IEEE Transactions on Power Systems, Vol. 11, No. 1, pp. 254-9, Feb. 1996.

[Abo91] Abouzahr, I., Ramakumar, R., "Loss of Power Supply Probability of Stand-Alone Photovoltaic Systems: A Closed Form Solution Approaches," IEEE Transactions on Energy Conversion, Vol. 1, No. 1, pp. 1-10, March 1991.

[Aga95] Aganagic, M., Mokhtari, S., "Coordination Between Short-Term and Long-Term Resource Scheduling," American Power Conference Proceedings, Vol.57-II, pp. 1719-1722, April 1995.

[Agu88] Agulár, R.J., Collares-Pereira, M., Conde, J.P., "Simple Procedure for Generating Sequence of Daily Radiation Values Using a Library of Markov Transition Matrices," Solar Energy, Vol. 40, pp. 269-279, 1988.

[Ahm98] Ahmad, A., Kothari, D.P., "A Review of Recent Advances in Generator Maintenance Scheduling," Electric Machines and Power Systems, Vol. 26, Iss. 4, pp. 373-87; May 1998.

[All81] Allan, R.N., Leite da Silva, A.M., Abu-Nasser, A.A., Burchett, R.C., "Discrete Convolution in Power System Reliability," IEEE Transactions on Reliability, Vol. R-30, No. 5, pp. 452-456, December 1981.

[Alo99] Alomoush, M., Shahidehpour, M., "Fixed Transmission Rights for Inter-Zonal and Intra-Zonal Congestion Management," IEE Proceedings, 1999.

[Alo00] Alomoush, M., Shahidehpour, M., "Generalized Model for Fixed Transmission Rights Auction," Electric Power Systems Research, 2000.

[Alo00] Alomoush, M., Shahidehpour, M., "Contingency Constrained Congestion Management with a Minimum Number of Adjustments in Preferred Schedules," International Journal of Electric Power and Energy Systems, 2000.

[Ara82] Araujo, G.L., Sanchez, E., Marti, M., "Determination of the Two-exponential Solar Cell Equation Parameters from Empirical Data," Solar Cells, Vol. 5, pp. 199-204, 1982.

[Aro 97] Arroyo, J.M., Conejo, A., Jimenez, N., "A Genetic Algorithm Approach to Solve the Unit Commitment Problem," ISAP' 97, PP. 182-187, July 1997.

[Bai96] Bai, X., Shahidehpour, M., "Hydro-thermal Scheduling by Tabu Search and Decomposition Method," IEEE Transactions on Power Systems, Vol. 11, No. 2, pp. 968-975, May 1996.

[Bai97a] Bai, X., Shahidehpour, M., "Extended Neighborhood Search Algorithm for Constrained Unit Commitment," International Journal of Electrical Power and Energy Systems, Vol. 19, No. 5, pp. 349-356, 1997.

[Bai97b] Bai, X., Shahidehpour, M., "Transmission Analysis by Nash Game Method," IEEE Transactions on Power Systems, Vol. 12, No. 3, pp. 1046-1053, August 1997.

[Bar84] Bartoli, B., Cuomo, V., Fontana, F., Serio, C., and Silvestrini, V., "The Design of Photovoltaic Plants: An Optimization Procedure," Applied Energy, Vol. 32, No. 1, pp. 37-47, 1984.

[Bat92] Batut, J., Renaud, A., "Daily Generation Scheduling Optimization With Transmission Constraints: A New Class of Algorithms," IEEE Transactions on Power Systems, Vol. 7, No.3, pp. 982-989, Aug. 1992.

[Bec77] Bechert, T.E., Chen, N., "Area Automatic Generation Control by Multi-Pass Dynamic Programming," IEEE Transactions on Power Apparatus and Systems, Vol. PAS-96, No. 5, pp. 1460-1469, Sept. 1977.

[Bey87] Beyer, H.G., Luther, J., Schumacher-Grohn, J., "Combined Battery/ Hydrogen Storage for Autonomous Wind/Solar Systems," Advances in Solar Energy Technology, Proceedings of the Biennal Congress of the International Solar Energy Society, pp. 422-425, Hamburg, 1987.

[Bil84] Billinton, R., Allan, R., "Reliability Evaluation of Power Systems," Plunum Press, New York, 1984.

[Blo83] Bloom, J., "Solving an Electricity Generating Capacity Expansion Planning Problem by Generalized Benders Decomposition," Operation Research, Vol. 31, No. 1, pp. 84-100, Jan./Feb. 1983.

[Boo72] Booth, R.R., "Power System Simulation Model Based on Probability Analysis," IEEE Transactions on Power Apparatus and Systems, Vol. PAS-91, pp. 62-69, 1972.

[Bre98] Bretthauer, G., Gamaleja, T., Handschin, E., Neumann, U., Koffmann, "Integrated maintenance scheduling system for electrical energy systems," IEEE Transactions on Power Delivery, Volume: 132, pp. 655 –660, April 1998.

[Cha85] Chalmer, M., et al, "The Effect of Photovoltaic Power Generation on Utility Operation," IEEE Transactions on Power Apparatus and Systems, Vol. PAS-104, No. 3, pp. 524-530, March 1985.

[Cha90] Chapman, R.N., "The Synthesis of Solar Radiation Data for Sizing Stand-alone Photovoltaic Systems," Twenty First IEEE Photovoltaic Specialist Conference, pp. 965-970, Orlando, Florida, 1990.

[Che91] Chen, L., Toyoda, J., "Optimal Generating Unit Maintenance Scheduling for Multi-area System with Network Constraints," IEEE Transactions on Power Systems, Vol. 6, pp. 1168-1174, 1991.

[Cho88] Chowdhury, B.H., Rahman, S., "Analysis of Interrelationships between Photovoltaic Power and Battery Storage for Electric Utility Load Management," IEEE Transactions on Power Systems, Vol. 3, No. 3, pp. 900-907, August 1988.

[Cho91] Chowdhury, B.H., "Photovoltaics-Utility Interconnection: An Integrated Approach," Proceedings of the Biennial Congress of the International Solar Energy Society, Vol. 1, pp. 143-148, August 1991.

[Chr72] Christiaanse, W.R., Palmer, A.H., "A Technique for the Automated Scheduling of the Maintenance of Generating Facilities," IEEE Transactions on Power Apparatus and Systems, Vol. PAS-91, No. 1, pp. 137-144, Jan./Feb. 1972.

[Chh82] Cohen, A.I., Yoshimura, M., "A Branch and Bound Algorithm for Unit Commitment," IEEE Transactions on Power Systems,Vol. PAS-101, No. 1, pp. 444-451, Jan. 1982.

[Coh87] Cohen, A.I., Wan, S.H., "A Method for Solving the Fuel Constrained Unit Commitment Problem," IEEE Transactions on Power Systems, Vol. No. 2, pp. 608-614, Aug. 1987.

[Dah97] Dahal, K.P., McDonald, J.R., "A Review of Generator Maintenance Scheduling Using Artificial Intelligence Techniques," 32nd Universities Power Engineering Conference, UPEC '97, Vol.2, pp. 787-90, UMIST, Manchester, UK, 1997.

[Day82] Day, J.T., Malone, M.J., "Electric Utility Modeling Extensions to Evaluate Solar Plants," IEEE Transactions on Power Apparatus and Systems, Vol. PAS-101, No. 1, pp. 120-126, January 1982.

[Dee90a] Deeb, N., Shahidehpour, M., "Improved System Operation by Optimal Adjustment of Reactive Generation in Electric Power Network," International Journal of Energy Research, Vol. 14, No. 1, pp. 651-664, March 1990.

[Dee90b] Deeb, N., Shahidehpour, M., "Linear Reactive Power Optimization in a Large Power Network Using the Decomposition

Approach," IEEE Transactions on Power Systems, Vol. 5, No. 2, pp. 428-438, May 1990.

[Dee90c] Deeb, N., Shahidehpour, M., "An Overview of Reactive Power Optimization in Electric Power Systems," Electric Machines and Power Systems, Vol. 18, No. 5, pp. 495-518, Sept. 1990.

[Dee90d] Deeb, N., Shahidehpour, M., "Cost Minimization in Allocation of Reactive Power Sources,' International Journal Electric Power and Energy Systems, Vol. 12, No. 4, pp. 263-270, Oct. 1990.

[Dee91] Deeb, N., Shahidehpour, M., "A Decomposition Approach for Minimizing Real Power Losses in Power Systems," IEE Proceedings, Part C, Vol. 138, No.1, pp. 27-38, Jan. 1991.

[Dee93] Deeb, N., Shahidehpour, M., "Cross Decomposition for Multi-Area Optimal Reactive Power Planning," IEEE Transactions on Power Systems, Vol. 8, No. 3, pp. 1539-1544, Nov. 1993.

[Deg73] DeGarmo, E.P., Canada, J.R., "Engineering Economy," Macmillan Company, New York, 1973.

[Del92] Delson, J., Shahidehpour, M., "Linear Programming Applications to Power System Economics, Planning and Operations," IEEE Transactions on Power Systems, Vol. 7, No. 3, pp. 1155-1163, Aug. 1992.

[Den92] Deng, Z., Singh, C., "A New Approach to Reliability Evaluation of Interconnected Power Systems Including Planned Outage and Frequency Calculations," IEEE Transactions on Power Systems, Vol. 7, No. 2, May 1992.

[Dhi81] Dhillon, B.S., Singh, C., "Engineering Reliability," John Wiley and Sons, New York, 1981.

[Dop75] Dopazo, J.F., Merrill, H.M., "Optimal Generator Maintenance Scheduling Using Integer Programming," IEEE Transactions on Power Apparatus and Systems, Vol. PAS-94, No. 5, pp. 1537-1545, Sept./Oct. 1975.

[Dop80] Dopazo, J.F.; Merrill, H.M., "Power Plant Maintenance Scheduling-a survey," Proceedings of the American Power Conference, Vol.42, pp. 412-17, Chicago, 1980.

[Elh79] El-Hawary, M.E., Christensen, G.S., "Optimal Economic Operation of Electric Power Systems," Academic Press, New York, 1979.

[End78] Endrenyi, J., "Reliability Modeling in Electric Power Systems," John Wiley and Sons, New York, 1978.

[Ega76] Egan, G.T., Dillon, T.S., Morsztyn, K., "An Experimental Method of Determination of Optimal Maintenance Schedules in Power Systems Using the Branch and Bound Techniques," IEEE Transactions on Man. and Cybernetics, Vol. SMC-6, No. 8, pp. 538-547, Aug. 1976.

[Esc82] Escudero, L.F., and Horton, J.W., Scheiderich, J.E., "On Maintenance Scheduling for Energy Generators," IEEE Transactions on Power Apparatus and Systems, Vol. PAS-101, No. 8, pp. 2770-2779, August 1982.

[Fed96] Federal Energy Regulatory Commission, "Capacity Reservation Open Access Transmission Tariffs," NOPR, April 1996.

[Fer96] Ferrero, R., Shahidehpour, M., "Energy Interchange in Deregulated Power Systems," International Journal of Electrical Power and Energy Systems, Vol. 18, No. 4, pp. 251-258, May 1996.

[Fer97a] Ferrero, R., Shahidehpour, M., "Dynamic Economic Dispatch in Deregulated Systems," International Journal of Electrical Power and Energy Systems, Vol. 19, No. 7, pp. 433-439, 1997.

[Fer97b] Ferrero, R., Shahidehpour, M., and Ramesh, V. "Transaction Analysis in Deregulate Power Systems using Game Theory," IEEE Transactions on Power Systems, Vol. 12, No. 3, pp. 1340-1347, August 1997.

[Fer97c] Ferrero, R., Shahidehpour, M., "Short-term Power Purchases Considering Uncertain Prices," IEE Proceedings, Vol. 144, No. 5, pp. 423-428, Sept. 1997.

[Fer98] Ferrero, R., Rivera, J., Shahidehpour, M., "Application of Games with Incomplete Information for Pricing Electricity in Deregulated Power Pools," IEEE Transactions on Power Systems, Vol. 13, No. 1, pp. 184-189, Feb. 1998.

[Fis79] Fischl, R., Herczfeld, P.R., Halprin, T.F., Stewart, D.B., "Design of Integrated-Electric-Solar-Utility System for Peak Load Shaving," Proceedings of the IEEE Power Engineering Soc.,

1979 Winter Meeting, New York, paper A 79 102-5, February 1979.

[Gar94] Gardner J., Hobs, W., Lee, F.N., Leslie, E., Streiffert, D., Todd, D., "Summary of the Panel Session 'Coordination between Short-Term Operation Scheduling and Annual Resource Allocation'," Proceedings of the IEEE Power Engineering Soc., 1994 Winter Meeting, paper 95 WM 154-5 PWRS, New York, 1994.

[Gar72] Garver, L.L., "Adjusting Maintenance Schedules to Levelize Risk," IEEE Transactions on Power Apparatus and Systems, Vol. PAS-91, No. 5, pp. 2057-2063, Sept./Oct. 1972.

[Geo72] A.M. Geoffrion, "Generalized Benders Decomposition," JOTA, Vol. 10, No. 4, pp. 237-261, 1972.

[Gro88] Gross, G., Garapic, N.V., McNutt, B., " The Mixture of Normal Approximation Technique for Equivalent Load Duration Curves," IEEE Transactions on Power Systems, Vol. 3, No. 2, pp. 368-374, May 1988.

[Gua91] Guan, X., Luh, P.B., Yan, H., Amalfi, J.A, "Short-term Scheduling of Thermal Power Systems," PICA Conference, pp. 120-126, May 1991.

[Hap94] Happ, H.H., "Cost of Wheeling Methodologies," IEEE Transactions on Power Systems, Vol. 9, No. 1, Feb. 1994.

[Har66] Hara, K., Kimura, M., Honda, N., "A Method for Planning Economic Unit Commitment and Maintenance of Thermal Power Systems," IEEE Transactions on Power Apparatus and Systems, Vol. PAS-85, No. 5, pp. 427-436, May 1966.

[Han92] Henley, E.J., Kumamoto, H., "Probabilistic Risk Assessment," IEEE Press, 1992.

[Hua92] Huang, J.C., Lin, E.C., "Fuzzy Approach for Generator Maintenance Scheduling," Electric Power System Research, No. 24, pp. 31-38, 1992.

[Hun98] Huang, S.J., "A Genetic-evolved Fuzzy System for Maintenance Scheduling of Generating Units," International Journal of Electrical Power & Energy Systems, Vol. 20, No. 3, pp. 191-195, 1998.

[Ili98] Illic, M., Galiana, F., and Fink, L., "Power System Restructuring: Engineering and Economics," Kluwer Academic Press, Boston, 1998.

[Kha86] Khalat, M.A., Rahman, S., "A Probabilistic Approach to Photovoltaic Generator Performance Prediction," IEEE Transactions on Energy Conversion, Vol. EC-1, No. 3, pp. 34-40, September 1986.

[Kha91] Al-Khamis, T.M., Vemuri, S. and Lemonidis, L., Yellen, J., "Unit Maintenance with Fuel Constraints," IEEE Transactions on Power Systems, PICA Proceedings, pp. 113-119, May 1991.

[Kha79] Khatib, H., "Maintenance Scheduling of Generating Facilities," IEEE Transactions on Power Apparatus and Systems, Vol. PAS-98, No. 5, pp. 1604-1608, September/October 1979.

[Kir96] Kirby, B., Hirst, E. "Unbundling Electricity: Ancillary Services," IEEE Power Engineering Review, pp. 5-6, June 1996.

[Kle87] Klein, S.A., Beckman, W.A., "Loss of Load Probabilities for Stand-Alone Photovoltaic Systems," Solar Energy, Vol. 39, No. 6, pp. 499-512, 1987.

[Kni91] Knight, K.M., Klein, S.A., Duffie, J.A., "Methodology for Synthesis of Hourly Weather Data," Solar Energy, Vol. 46, pp. 109-120, 1991.

[Kra88] Kralj, B.L., Petrovic, R., "Optimal Preventive Maintenance Scheduling of Thermal Generating Units in Power Systems- A survey of Problem Formulations and Solution Methods," European Journal of Operational Research, Vol. 35, Iss. 1, pp. 1-15, April 1988.

[Kum86] Kumar, A.B.R., Vemuri, S., Ebrahimzadeh, P., Farahbakhshian, N., "Fuel Resource Scheduling- The Long Term Problem," IEEE Transactions on Power Systems, Vol. PWRS-1, No. 4, 1986.

[Las70] Lasdon, L., Optimization Theory for Large Systems, New York, MacMillan, 1970.

[Lee91] Lee, F.N., "The Coordination of Multiple Constrained Fuels," IEEE Transactions on Power Systems, Vol. 6, No. 2, pp. 699-707, May 1991.

[Lee92a] Lee, T.Y., Chen, N., "The Effect of Pumped Storage and Battery Energy Storage Systems on Hydrothermal Generation Coordination" IEEE Transactions on Energy Conversion, Vol. 7, No. 4, pp. 631-637, Dec. 1992.

[Lee92b] Lee, F.N., Liao, J., Breiphol, A., "Adaptive Fuel Allocation Using Pseudo Fuel Prices," IEEE Transactions on Power Systems, Vol. 7, No. 2, pp. 487-496, May 1992.

[Li93] Li, S., Shahidehpour, M. .and Wang, C., "Promoting the Application of Expert Systems in Short-Term Unit Commitment," IEEE Transactions on Power Systems, Vol. 8, No. 1, pp. 286-292, Feb. 1993.

[Liu63] Liu, B.Y.H., Jordan, R.C., "The Interrelationship and Characteristic Distribution of Direct, Diffuse and Total Solar Radiation," Solar Energy, Vol. 7, No. 2, pp. 53-74, 1963.

[Ma97a] Ma, H., Shahidehpour, M., "Decomposition Approach to Unit Commitment with Reactive Constraints," IEE Proceedings, Vol. 144, No. 2, pp. 113-117, March 1997.

[Ma97b] Ma, H., Shahidehpour, M., "Unit Commitment with Voltage Constraints," International Journal of Energy Research, Vol. 21, pp. 1385-1398, Nov. 1997.

[Ma98] Ma, H., Shahidehpour, M., "Transmission Constrained Unit Commitment Based on Benders Decomposition," Electric Power and Energy Systems, Vol. 20, No. 4, pp. 287-294, April. 1998.

[Ma99] Ma, H., Shahidehpour, M., "Unit Commitment with Transmission Security and Voltage Constraints," IEEE Transactions on Power Systems, Vol. 14, No. 2, May 1999.

[Mar96] Marwali, M., Rezania, E., Shahidehpour, M., "Photovoltaic Stand-alone Simulation Method for Rural Electrification," Energy, Electrical Power Systems, Water Resources, and Agriculture Proceedings, pp. 117-126, The City University of New York, NY, Sept. 1996.

[Mar97a] Marwali, M., Shahidehpour, M., Ramesh, V.C., "A Decomposition Approach to Generation Maintenance Scheduling with Network Constraints," American Power Conference Proceedings, Vol.59-I, pp. 110-115, April 1997.

[Mar97b] Marwali, M., Shahidehpour, M., Daneshdoost, M., "Probabilistic Production Costing for Renewable Energy in Utility with Battery Storage," IEEE Transaction on Energy Conversion, Vol. 4, No. 2, pp. 175-180, June 1997.

[Mar98a] Marwali, M., Shahidehpour, M., "Integrated Generation and Transmission Maintenance Scheduling with Network Constraints," IEEE Transaction on Power Systems, Vol. 13, No. 3, pp. 1063-1068, August 1998.

[Mar98b] Marwali, M., Ma, H., Shahidehpour, M., Abdul-Rahman, K.H., "Short Term Generation Scheduling in Photovoltaic-Utility Grid with Battery Storage," IEEE Transaction on Power Systems, Vol. 13, No. 3, pp. 1057-1062, August 1998.

[Mar98c] Marwali, M., Shahidehpour, M., "A Deterministic Approach to Generation and Transmission Maintenance Scheduling with Network Constraints," Electric Power Systems Research, Vol. 47, No. 2, pp. 101-114, Oct. 1998.

[Mar99a] Marwali, M., Shahidehpour, M., "Long-Term Transmission and Generation Maintenance Scheduling with Network, Fuel, and Emission Constraints," IEEE Transaction on Power Systems, Vol.14, pp.1160-1165, Aug. 1999.

[Mar99b] Marwali, M., Shahidehpour, M., "Short-term Transmission Line Maintenance Scheduling in a Deregulated System," PICA '99, pp. 31 –37, California, 1999.

[Mar00] Marwali, M., Shahidehpour, M., "Coordination between Long-Term and Short-Term Generation Scheduling in a Deregulated System," IEEE Transactions on Power Systems, 2000.

[May88] Mayer, P.G., Mayer, I., "On Calculating the Position of the Sun," Int. Journal of Ambient Energy, Vol. 9, No. 1, pp. 47-51, 1988.

[Men88] Menicucci, D.F., Fernandez, J.P., "User's Manual for PVFORM: A Photovoltaic System Simulation Program for Stand-Alone and Grid-Interactive Applications," Sandia Report SAND85-0376.UC-276, Albuquerque, NM. 1988.

[Mic88] Michalsky, J.J., "The Astronomical Almanac's Algorithm for Approximate Solar Position (1950-2050)," Solar Energy, Vol. 40, No. 3, pp. 227-235, 1988.

[Mok88] Mokhtari, S, Singh, J., Wollenberg, B., "A Unit Commitment Expert System," IEEE Transactions on Power Systems, Vol. 3, No. 1, pp. 272 – 277, Feb. 1988.

[Mon87] Monticelli, A., Pereira, M.V.F., Granville, S., "Security-Constrained Optimal Power Flow with Post-Contingency Corrective Rescheduling," IEEE Transactions on Power Systems, Vol. 2, No. 2, pp.175-182, Feb. 1987.

[Mos91] Moslehi, K., Sherkat, V.R., Ruiz, F.C., "Optimal Scheduling of Long-term Fuel Purchase, Distribution, Storage and Consumption," IEEE Transactions on Power Systems, PICA Proceedings, pp. 98-104, May 1991.

[Ner96] National Electric Reliability Council, "Available Transfer Capability Definitions and Determination," NERC Report, Washington D.C., May 1996.

[Ofr83] Ofry, E., Braunstein, A., "The Loss of Power Supply Probability as a Technique for Designing Stand-Alone Solar Electrical (Photovoltaic) Systems," IEEE Transactions on Power Apparatus and Systems, Vol. PAS-102, No. 5, pp. 1171-1175, May 1983.

[Oli89] Oliveira, G.C., Pereira, M.V.F., Cunha, S.H.F., "A Technique for Reducing Compuational Effort in Monte-Carlo Based Composite Reliability Evaluation," IEEE Transactions on Power Systems, Vol 4, No.4 , pp.1309-1315, October 1989.

[Org77] Orgill, J.F., Hollands, K.G.T., "Correlation Equation for Hourly Diffuse Radiation on A Horizontal Surface," Solar Energy, Vol. 19, No. 4, pp. 357-359, 1977.

[Ouy90] Ouyang, Z., Shahidehpour, M., "Short-Term Unit Commitment Expert System," Electric Power Systems Research, Vol. 20, No. 1, pp. 1-13, Nov. 1990.

[Ouy91a] Ouyang, Z., Shahidehpour, M., "A Heuristic Multi-Area Unit Commitment with Economic Dispatch," IEE Proceedings, Part C, Vol. 138, No. 3, pp. 242-252, May 1991.

[Ouy91b] Ouyang, Z., Shahidehpour, M., "An Intelligent Dynamic Programming for Unit Commitment Applications," IEEE Transactions on Power Systems, Vol. 6, No. 3, pp. 1203-1209, Aug. 1991.

[Ouy91c] Ouyang, Z., Shahidehpour, M., "A Heuristic Short-Term Unit Commitment," IEEE Transactions on Power Systems, Vol. 6, No. 3, pp. 1210-1216, Aug. 1991.

[Ouy92a] Ouyang, Z., Shahidehpour, M., "A Hybrid Artificial Neural Networks-Dynamic Programming Approach to Unit Commitment," IEEE Transactions on Power Systems, Vol. 7, No. 1, pp. 236-242, Feb. 1992.

[Ouy92b] Ouyang, Z., Shahidehpour, M., "A Multi-Stage Intelligent System for Unit Commitment," IEEE Transactions on Power Systems, Vol. 7, No. 2, pp. 639-646, May 1992.

[Per85] Pereira, M.V.F., Pinto, L.M.V.G, Cuncha, S.H.F., Oliveira, G.C., "Decomposition Approach to Automated Generation /Transmission Expansion Planning," IEEE Transactions on Power Apparatus and Systems, Vol. PAS-104, No. 11, pp. 3074-3080, Nov. 1985.

[Qiu87] Qiu, J., and Shahidehpour, M., "A New Approach for Minimizing Power Losses and Improving Voltage Profile," IEEE Transactions on Power Systems, Vol. PWRS-2, No. 2, pp. 287-295, May 1987.

[Rau80] Rau, N.S., Toy, P., and Schenk, K.F., "Expected Energy Production Costs by the Method of Moments," IEEE Transactions on Power Apparatus and Systems, Vol. PAS-99, No. 5, pp.1908-1915, Sept./Oct. 1980.

[Rel79] Reliability Test System Task Force of the Application of Probability Methods Subcommittee, "IEEE Reliability Test System," IEEE Transactions on Power Apparatus and Systems, Vol. PAS-98, No. 6, pp. 2047-2054, Nov./Dec. 1979.

[Rez99] Rezania, E., Shahidehpour, M., "Calculation of Transfer Capability of Power Systems using an Efficient Predictor-Corrector Primal-Dual Interior-Point Algorithm," Electric Machines and Power Systems, Vol. 27, No. 1, pp. 25-38, Jan.1999.

[Ruz91] Ruzic, S., Rajakovic, N., "A New Approach for Solving Extended Unit Commitment Problem," IEEE Transactions on Power Systems, Vol. 6, No.1, pp.269-277, Feb. 1991.

[Saa99] Saadat, H., "Power System Analysis," McGraw-Hill, New York, 1999.

[Sat91] Satoh, T., Nara, K. "Maintenance Scheduling by Using Simulated Annealing Method," IEEE Transactions on Power Systems, Vol. 6, No.2, May 1991.

[Sau98] Sauer, P.W., Pai, M.A., "Power System Dynamics and Stability," Prentice Hall, New Jersey, 1998.

[Sch84] Schenk, K.F., Misra, R.B., Vassos, S., Wen, W., "A New Method for the Evaluation of Expected Energy Generation and Loss of Load Probability," IEEE Transactions on Power Apparatus and Systems, Vol. PAS-103, No. 2, pp. 294-303, February 1984.

[Sha99] Shahidehpour, M., Yamin, H.Y., "Risk Management using Game Theory in Transmission Constrained Unit Commitment within a Deregulated Power Market," IEEE/PES Tutorial on Applications of Gaming Methods to Power System Operation, 1999.

[Shb99] Sheblé, G.B., "Computation Auction Methods for Restructured Electric Power Industry," Kluwer Academic Press, Boston, 1999.

[She65] Shepherd, C.M., "An Equation Describing Battery Discharge," Journal Electrochem. Soc., Vol. 112, pp. 657-664, 1965.

[Shi89] Shirmohammadi, D., et al, "Evaluation of Transmission Network Capacity Use for Wheeling Transaction," IEEE Transactions on Power Systems, Vol. 4, No. 4, pp. 1405-1413, October. 1989.

[Shi91] Shirmohammadi, D., Rajagopalan, C., Alward, E.R., Thomas, C.L., "Cost of Transmission Transactions: Introduction," IEEE Transactions on Power Systems, Vol. 6, No. 3, pp. 1006-1016, August 1991.

[Shi96] Shirmohammadi, D., Filho, X.V., Gorenstin, B., Pereira, M.V.P., "Some Fundamental Technical Concepts About Cost Based Transmission Pricing," IEEE Transactions on Power Systems, Vol. 11, No. 2, May 1996.

[Sil95] Silva, E.L., Morozowski, M., Fonseca, L.G.S., Oliveira, G.C., Melo, A.C.G., Melo, J.C.O., "Transmission Constrained Maintenance Scheduling of Generating Units: A Stochastic Programming Approach," IEEE Transactions on Power Systems, Vol. 10, No.2, pp. 695-701, May 1995.

[Sin99] Singh, H., Papalexopoulos, A., "Competitive Procurement of Ancillary Services by an Independent System Operator," IEEE Transactions on Power Systems, Vol. 14, No. 2, pp. 498-504, May 1999.

[Sin99] Song, Y.H., "Modern Optimization Techniques in Power Systems," Kluwer Academic Publishers, Boston, 1999.

[Spe65] Spencer, J.W., "Fourier series representation on the Position of the Sun," Search, Vol. 2, No. 5, pp. 172-201, 1965.

[Str80] Stremel, J.P., Jenkins, R.T., Babb, R.A., and Bayless, W.D., "Production Costing Using the Cumulant Method of Representing the Equivalent Load Curve," IEEE Transactions on Power Apparatus and Systems, Vol. PAS-99, No. 5 pp. 1947-1956, Sept./Oct. 1980.

[Szk99] Szkuta, B.R., Sanabria, L.A., and Dillon, T.S., "Electricity Price Short-Term Forecasting using Artificial Neural Networks," IEEE Transactions on Power Systems, Vol. 14, No. 3, pp. 851-857, August 1999.

[Ton89] Tong, S., Shahidehpour, M., "Combination of Lagrangian Relaxation and Linear Programming Approaches for Fuel Constrained Unit Commitment Problem," IEE Proceedings, Part C, Vol. 136, No. 3, pp. 162-174, May 1989.

[Ton91a] Tong, S., Shahidehpour, M., and Ouyang, "Heuristic Short-term Unit Commitment," IEEE Transactions on Power Systems, Vol. 6, No. 3, pp. 1210-1216, Aug. 1991.

[Ton91b] Tong, S., Shahidehpour, M., "An Overview of Power Generation Scheduling in the Optimal Operation of a Large Scale Power System," Electric Machines and Power Systems, Vol. 19, No. 5, PP. 731-762, Sept. 1991.

[Ton92] Tong, S., Shahidehpour, M., "A Scheduling Model for the Daily Operation of an Electric Power System," Applied Mathematical Modelling, Vol. 16, pp. 226-244, May 1992.

[Vic94] Vickers, V.L., Hobbs, W.J., Vemuri, S., Todd, D.L., "Fuel Resources Scheduling with Emission Constraints," IEEE Transactions on Power Systems, Vol. 9, No. 3, pp. 1531-1538, August 1994.

[Vem92] Vemuri, S., Lemonidis, L., "Fuel Constrained Unit Commitment," IEEE Transactions on Power Systems, Vol. 7, No. 1, pp.410-415, Feb. 1992.

[Wan88] Wang, X., "Equivalent Energy Function Approaches to Power System Probabilistic Modeling," IEEE Transactions on Power Systems, Vol. 3, No. 3, pp. 823-829, August 1988.

[Wan92] Wang, C., Shahidehpour, M., "A Decomposition Approach to Non-Linear Multi-Area Generation Scheduling with Tie-Line Constraints using Expert Systems," IEEE Transactions on Power Systems, Vol. 7, No. 3, pp. 1409-1418, Nov. 1992.

[Wan93] Wang, C., Shahidehpour, M., "Effect of Ramp Rate Limits on Unit Commitment and Economic Dispatch", IEEE Transactions on Power Systems, Vol. 8, No. 3, pp. 1341-1350, Aug. 1993.

[Wan94] Wang, C., Shahidehpour, M., "Ramp-Rate Limits in Unit Commitment and Economic Dispatch Incorporating Rotor Fatigue Effect," IEEE Transactions on Power Systems, Vol. 9, No. 3, pp. 1539-1545, Aug. 1994.

[Wan95a] Wang, C., Shahidehpour, M., "Optimal Generation Scheduling with Ramping Costs," IEEE Transactions on Power Systems, Vol. 10, No. 1, pp. 60-67, Feb. 1995.

[Wan95b] Wang, S., Shahidehpour, M., and Mokhtari, S., "Short-term Generation Scheduling with Transmission and Environmental Constraints Using an Augmented Lagrangian Relaxation," IEEE Transactions on Power Systems, Vol. 10, No. 3, pp. 1294-1301, Aug. 1995.

[Wan95c] Wang, S., Shahidehpour, M., "Power Systems Marginal Cost Curve and Its Applications," IEEE Transactions on Power Systems, Vol. 10, No. 3, pp. 1321-1328, Aug. 1995.

[Wan98] Wang, Y., Handschin, E., "Unit Maintenance Scheduling in Open Access Electric Power Systems," IEEE Power Engineering Review, Vol.18, No.12, pp. 64-5,Dec. 1998.

[Wan99] Wang, Y., Handschin, E., "Unit Maintenance Scheduling in Open Systems Using Genetic Algorithm, IEEE Transmission and Distribution Conference, Vol. 1, pp. 334 –339, 1999.

[Wan00] Wang, Y., Handschin, E., "A New Genetic Algorithm for Unit Maintenance Scheduling of Power Systems," International

Journal of Electrical Power & Energy Systems, Vol. 22, No. 5, pp. 343-348, May 2000

[Was96] University of Washington, "IEEE 118-bus system," available at the following web site: http://www/wahoo.ee.washington.edu , 1996.

[Web01] Web site: http://WWW.caiso.com: Sponsored by the California Independent System Operator.

[Web02] Web site: http://WWW.calpx.com: Sponsored by the California Power Exchange.

[Web03] Web site: http://WWW.ornl.com: Sponsored by Oak Ridge National Laboratory.

[Web04] Web site: http://WWW.pjm.com: Sponsored by Pennsylvania-NewJersy-Maryland interconnection.

[Web05] Web site: http://WWW.iso-ne.com: Sponsored by New England Inc.

[Web06] Web site: http://WWW.ferc.fed.us: Sponsored by Federal Energy Regulation Commission.

[Web07] Web site: http://WWW.eroc.som: Sponsored by Division of Electric Reliability Council of Texas.

[Web08] Web site: http://WWW.midwestiso.org: Sponsored by Midwest Independent Transmission System Operator, LLC.

[Web09] Web site: http://WWW.nerc.com: Sponsored by North America Electric reliability Council.

[Web10] Web site: http://WWW.nypowerpool.com: Sponsored by New York Independent System Operator.

[Web11] Web site: http://WWW.idaopower.com: Sponsored by Independent Grid Operator.

[Web12] Web site: http://WWW.cnie.com: Sponsored by the Committee for the National Institute for environment.

[Web13] Web site http://www.epis.com: Sponsored by Electric Price Information and Services Inc.

[Wet89] Wets, R. J-B, "Scenario Analysis vs. Stochastic Optimization," Proceedings of the Workshop on Resource Planning Under Uncertainty for Electric Power Systems, Stanford University, November 1989.

[Woo96] Wood, A.J., Wollenberg, B.F., "Power Generation, Operation & Control," 2nd edition, John Wiley and Sons, New York, 1996.

[Yam81] Yamayee, Z.A., Peschon, J., "Utility Integration Issues of Residential Photovoltaic Systems," IEEE Transactions on Power Apparatus and Systems, Vol. PAS-100, No. 5, pp. 2365-2373, May 1981.

[Yam82] Yamayee, Z.A., "Maintenance Scheduling: Description, Literature Survey, and Interface with Overall Operations Scheduling.," IEEE Transactions on Power Apparatus and Systems, Vol. PAS-101, Iss. 8, pp. 2770-9, Aug. 1982.

[Yam99a] Yamin, H.Y., Shahidehpour, M., "Competitive Energy and Ancillary Service Markets," Proceedings of the IEEE 31st North American Power System Symposium (NAPS), pp. 10-16, Oct. 1999.

[Yam99b] Yamin, H.Y., Shahidehpour, M., "Electricity Pricing Utilizing Transmission Constrained Unit Commitment and Game Theory in A Deregulated Power Marketplace," 32nd Hawaii International Conference on System Sciences, HICSS-32, Jan. 1999.

[Yamweb] Yamin, H.Y., Shahidehpour, M., "IEEE 118-bus system," available at the following web site: http://www.iit.edu/~yamihat/

[Yel92] Yellen, J., Al-Khamis, T.M., Vemuri, S., Lemonidis, L., "A Decomposition Approach to Unit Maintenance Scheduling," IEEE Transactions on Power Systems, Vol. 7, No. 2, pp. 726-733, May 1992.

[Yos93] Yoshimoto, K., Yasuda, K., Yokoyama, R., Tanaka, H., Akimoto, Y., "Extended Hopfield Neural Network Applied to Maintenance Scheduling of Generating Units Based on Decomposition," ISAP '93, Australia, 1993.

[Yu97] Yu, C.W., David, A.K., "Pricing Transmission Services in Context of Industry Deregulation," IEEE Transactions on Power Systems, Vol. 12, No. 1, May 1997.

[Zur75] Zurn, H.H., Quintana, V.H., "Generator Maintenance Scheduling via Successive Approximation Dynamic Programming," IEEE Transactions on Power Apparatus and Systems, Vol. PAS-94, No. 1, pp. 665-671, March/April 1975.

[Zur77] Zurn, H.H., Quintana, V.H., "Several Objective Criteria for Optimal Generator Preventive Maintenance Scheduling," IEEE Transactions on Power Apparatus and Systems, Vol. PAS-96, No. 3, pp. 984-992, May/June 1977.

INDEX

E

F

G

H

I

J

K

L

M

N

O

P

Q

R

S

T

U

V

W

Z

Lightning Source UK Ltd.
Milton Keynes UK
23 August 2010

158866UK00004B/36/A

9 780792 378723